17.00

Cover illustration: The cover picture shows the molecular and supermolecular structure of protein fibers as proposed by L. Pauling in 1950. The protein molecules were assumed to be in the conformation of a left-handed alpha-helix. Six protein molecules each are wound around a central protein molecule, forming a seven-stranded cable. The drawing is based on a picture in L. Pauling's „Chemie – Eine Einführung", Verlag Chemie, Weinheim 1956.

Fiber proteins are the main components of our hair, fingernails, and muscles. We know today that each protein fiber has a different structure. Wool and silk, discussed in this book, are examples.

Hans-Georg Elias

Mega Molecules

Tales of Adhesives, Bread, Diamonds, Eggs,
Fibers, Foams, Gelatin, Leather, Meat, Plastics,
Resists, Rubber, . . . and Cabbages and Kings

With 55 Figures and 34 Tables

Springer-Verlag
Berlin Heidelberg New York
London Paris Tokyo

Professor Dr. Hans-Georg Elias
4009 Linden Drive
Midland, MI 48640/USA

Title of the Original German Edition: Große Moleküle
© Springer-Verlag Berlin Heidelberg 1985
ISBN 3-540-15599-6; 0-387-15599-6

ISBN 3-540-17541-5 Springer-Verlag Berlin Heidelberg New York
ISBN 0-387-17541-5 Springer-Verlag New York Berlin Heidelberg

Library of Congress Cataloging-in-Publication Data
Elias, Hans-Georg, 1928-. Mega molecules. Translation of: Grosse Moleküle.
Bibliography: p. Includes index. 1. Macromolecules. I. Title.
QD381.E4313   1987   547.7   87-4278

© Springer-Verlag Berlin Heidelberg 1987
Printed in Germany

Typesetting, printing, and binding: Appl, Wemding
2152-3140-543210

. . . And furthermore, my son, be admonished:
of making many books there is no end; . . .

Ecclesiastes 12,12

Preface

All life is based on big molecules, scientifically called "macromolecules". Humans, animals, and plants cease to exist without these structural, reserve, and transport molecules. No life can be propagated without macromolecular DNA and RNA. Without macromolecules, we would only dine on water, sugars, fats, vitamins and salts but had to relinquish meat, eggs, cereals, vegetables, and fruits. We would not live in houses since wood and many stones consist of macromolecules. Without macromolecules, no clothes since all fibers are made from macromolecules. No present-day car could run: All tires are based on macromolecules. Without macromolecules no photographic films, no electronics ...

If macromolecules are so important then why is commonly so little known about their roles and why are they so little mentioned in school, if at all? As often in human history, tradition is important and science makes no exception. Chemistry was established as the chemistry of low molecular weight compounds since these were most easy to investigate, characterize, and convert. A beautiful tower of thought was erected by the chemical sciences long before the idea of giant molecules, macromolecules, took hold. There was no space for newcomers in this tower. Even today one can learn about chemistry without hearing a word about macromolecules.

There may be another reason. Serious students of macromolecules have to make frequent excursions into neighboring disciplines: Chemistry, physics, biology, materials science, and engineering. Rarely does another scientific subdiscipline lead to easily from the chemical structure of molecules to the physical structure of their assemblies and from there to properties which we see around us every day. It is exactly this connection with the useful which makes macromolecular science deeply suspect to the purists among scientists. Was not Aristotle the first to claim that a science has to be ranked the higher the less useful it was ...

This booklet thus wants to lead from the experiences of daily life to the ideas about structure and function of macromolecular compounds. Properties of glues, plastics, multigrade engine oils, rubbers, foams etc. will be traced back to their chemical and physical structures. The hardening of modern glues and the sweetening of old potatoes will be given as examples of chemical reactions of macromolecules, the staling of bread and the ironing of fabrics as examples of physical transitions.

The subject will not be rigorously treated though. This booklet was not intended to be a textbook but it should provide an easy access to the fascinating world of giant molecules. Those who have not forgotten to ask "Why?" should be especially interested and so should be teachers of all grades, scientists, and engineers. I would be pleased if I can convince politicians and educators to pay more attention to this discipline since macromolecules form the basis for all necessities of our material life: Food, clothing, shelter, transportation, and health.

The book is based on a preceding German edition which was subsequently translated, updated, modified, and yes, "Americanized" in content and style. I appreciate the help of Dr. Dale J. Meier, Professor of Polymer Physics at Michigan Molecular Institute, who read the draft and made numerous suggestions for improvements. The remaining mistakes are, of course, all mine.

Oh, and why "Mega Molecules" as title? Well, there are many reasons: the nice alliteration[1], the easy pronounciation[2], the simple spelling[3], the scientific correctness[4], the previous precedents[5], ...

Midland, Michigan   Hans-G. Elias
Fall 1986

---

[1] Q.e.d.
[2] Who can pronounce "macromolecule"?
[3] The author has quite a collection of misspellings of "macromolecular", he likes "mackarel" best.
[4] mega = 1 million (see Appendix). The better macromolecules have molar masses of about 1 million.
[5] If "Mega Trends", why not "Mega Molecules"?

Acknowledgements

The author appreciates the permission to reproduce figures and other material by the following publishers and individuals:

Academic Press, London (Fig. 15, from D. Lang, H. Bujard, B. Wolff, D. Russell, Journal of Molecular Biology, Volume *23* (1967) page 163)

Dr. D. T. Clarke, The Dow Chemical Company, Midland, Michigan (Fig. 39)

Hüthig and Wepf Publishers, Basle, Switzerland (Fig. 48; from H.-G. Elias, Makromoleküle, Fourth Edition 1981)

Institute for Applied Microscopy of the Fraunhofer Society, Karlsruhe, Federal Republic of Germany (Figs. 32 and 34)

A. Knopf Publisher, New York (Motto to Chapter 10; from J. Updike, Midpoint and Other Poems, 1968)

Professor Christian Ruscher, Department of Polymer Chemistry, Academy of Sciences of the German Democratic Republic, Teltow-Seehof, GDR (Fig. 31)

Verlag Lorenz, Vienna, Austria (Fig. 42; from J. Koppelmann, Oesterreichische Kunststoff-Zeitung 1981)

Table of Contents

1 Genuine Plastics and Other Natural Products

*"When I use a word," Humpty Dumpty said, in
rather a scornful tone, "it means just what
I choose it to mean – neither more nor less."
"The question is," said Alice, "whether you
can make words mean so many different things."
"The question is," said Humpty Dumpty, "which
is to be master – that's all."*

Lewis Carroll, Through the Looking Glass

Unknown are the reasons why early man searched for useful materials.
Was it sheer necessity, such as better protection against the inclemencies of
the weather, was it accidental observations, or just the impulse to toy
around? Man took what was available: Wood for lumber and fire, stones
for tools and construction, animal and vegetable fibers for clothing, amber
for jewelry, asphalt as adhesive. Next came the tinkerers who wanted to
improve old things or create new ones, not very systematically though and
often with wrong ideas, but occasionally successful. And finally science,
trying to explain the multitude of phenomena and to order them in systems with less and less assumptions, and then technology which converts
these insights into useful applications.

Long past are the times in which natural materials provided directly all
the needs of mankind, much longer than people commonly perceive. Man
tried very early to produce materials synthetically. Some of these synthetic
materials are so old that many people consider them as "natural". Elemental copper is a case in point. Long used before the Bronze Age, it is found
as genuine metal in large amounts only near Lake Superior; practically all
copper metal is however produced synthetically, i. e. chemically, from copper ores which are chemical compounds of copper with other chemical
elements. Bronze is produced synthetically from copper and tin; neither
bronze nor tin exist as such in nature. Elemental iron can be found in the
earth crust but only in very small amounts dispersed in basalt. The Iron
Age became a reality after our ancestors invented a chemical process for
making iron metal from iron ore. Glass is found occasionally in lava, all
our glass products, however, are synthetic. Ironstone and porcelain do not
exist as such in nature. Hides are converted into leather by tanning, in
most cases as a result of a chemical reaction. Copper, bronze, iron, glass,

ironstone, porcelain, and leather are thus not natural products. They are "synthetics", but are not called that for historical reasons.

Nostalgic people feel nonetheless that these old "synthetics" are somehow more natural and more compatible with man than the modern "true" synthetics. They seem to perceive that these old materials, often already mentioned by the bible, were taylormade by our good Lord for the direct utilization by present mankind. But cotton obviously does not grow to serve in ladies' underwear. Silk worms do not die happier in order to give beautiful silk ties. The Lord's plan for creation certainly did not consider the biosynthesis of natural rubber as a means for producing rubber tires. Mankind rather adopted these materials since nothing was available for the desired purposes. Certain applications require however special materials whether sanctified by tradition or not.

Mankind has a rather high demand for materials as Table 1 shows for the U.S. consumption per capita. The average American citizen uses per year about 5900 kg (12976 lbs) of sand, gravel and stones, and over 508 kg (1118 lbs) of wood and wood products, not including fire wood. Most of the materials mentioned in Table 1 are made up by macromolecules, the topic of this booklet: Plastics, resins, rubbers, fibers, wood, and leather, even cement, clays, and some stones, sands, and gravel, although the latter ones are often not recognized as genuine macromolecules.

Consumption is defined as the sum of production and imports minus exports; stockpiling is not considered. Data marked with an asterisk are production data for which no export/import figures are known. Net import is the difference between imports and exports. Data in parentheses are net exports. – Data are given in the internationally and scientifically used physical unit of kilogram, not the traditionally used pound (1 kg ≈ 2.2 lbs).

Table 1 also shows that the United States depends heavily on imports for some materials: 15% of the raw steel, 31% of the copper, 69% of the zinc, 66% of the semi-synthetic fibers, 92% of wool and silk, and 100% of natural rubber are (net) imported. On the other hand, net exports exist for cotton (50% of the production), synthetic rubbers (12%) and plastics and resins (9%). These data do not include materials used in finished and semi-finished goods such as textiles from Korea or rubber tires on Japanese cars.

No, it will not be possible to replace all modern synthetic materials by old "natural" ones as some well-meaning but misinformed "naturalists" propagate. What means "natural" anyway? Are polystyrene, polyethylene, and polymethyl methacrylate, all used as plastics, unnatural since they are produced by man from petroleum? Isn't petroleum also a natural product? Are you surprised when I tell you that polystyrene, polyethylene, and

Table 1. Per capita consumption of selected materials in the United States in 1983 (Estimated population: 234 000 000)

Materials	Consumption in kg per capita			Net import
	Natural products	Physically or chemically modified natural products	Synthetic products	in % of consumption
Non-renewable materials				
Non-metallic minerals				
Sand, gravel	2576			0
Stones	3322			0
Cement		287		6
Clays		270		0
Gypsum	81			39
Metals				
Raw steel			421	15
Aluminum, primary			15.0	6
Copper			6.4	31
Lead			4.8	10
Zinc			3.8	69
Polymers				
Plastics, resins			71	(9)
Synthetic rubbers			6.7	(12)
Synthetic fibers			16.9*	?
Total non-renewables	5979	557	546	
Renewable materials (all based on macromolecules)				
Wood and wood products				
Lumber	166			22
Plywood		29.6		4
Paper		277		6
Other	35			?
Fibers				
Cotton	5.5			(50)
Rayon, acetate		3.6		66
Wool, silk	1.6			92
Leather		6.4*		?
Natural rubber		2.6		100
Plastics, resins		0.2*		?
Total renewables	208	319	0	
Total non-renewables	5979	557	546	
Grand total	6187	876	546	

polymethyl methacrylate are also natural products, that they can be found as such in nature?

The old materials, as a rule, require considerable land areas for their production and as Will Rogers once remarked "They don't make the stuff any more". Mankind thus has the choice to use the available land area either for the production of food or for the manufacture of materials. On average, one can produce per 10000 square meters of land (about 2.5 acres)

        30 kg of wool or
       500 kg of natural rubber or
     1 000 kg of cotton or
     7 000 kg of wheat or
    10 000 kg of rice or
    70 000 kg of sugar beets or
 9 000 000 kg of synthetic fibers.

As far as I am concerned: I prefer natural food and "synthetic" materials over "natural" materials and synthetic foods. But that is really a matter of taste . . .

Both natural foods and natural materials we cannot have. Sure, highly industrialized countries can waste less and consume less but that is a matter of cost and would probably also mean more unemployment. It would also be unjust to demand from the people in less developed countries to forget about a better life which means not so much higher spiritual values than better quality of life, that is more and better food, clothing, housing, transportation, consumer goods and so on.

Even if we decide to use from now on only the old materials: We do not have the necessary land areas required for the present consumption, even if we disregard the additional desires by developing countries. Experts estimate that natural rubber can provide only 50% of the present world consumption of elastomers; natural rubber cannot be used for all modern applications, however. Additional 170000 square kilometers (66400 square miles) of best farm land would be necessary to replace by cotton the 13 million tons of synthetic fibers produced worldwide, i.e., an area about the size of the state of Missouri or twice the size of Austria or four times the size of Switzerland.

An increased production of natural fibers or leather would also do little for environmental protection. Most wool is produced by sheep and goats; both animals are prime destroyers of vegetation, as the Mediterranean countries show. Metals and glass require for their production much more oil than plastics and thus use much more of a limited resource. The manufacture of 1 million square meters (11 million square feet) of packaging film from polypropylene requires 76 tons of oil but the same area of cellophane (from wood) requires 178 tons. We need 66 tons of oil to make one million 1-Liter bottles from polyvinyl chloride (PVC) yet must use 230 tons to fabricate the same number of bottles from glass. One hundred kilometers (62.5 miles) of drainage pipes with 10 cm (4 inch) diameter require 1970 tons of oil if they are made from cast iron but only 275 tons when PVC is used. More and more of the old energy-consuming materials are thus being replaced by the energy-saving new synthetic materials.

2 In the Beginning was the Deed

. . . and he who seeks finds . . .

Matthew 7,8

2.1 From Rubber to Elastomers

Chance observations, purposeless toying and systematic tinkering often precede the scientific exploration of phenomena. Such was also the case with the giant molecules, the building elements of rubber, fibers, and plastics:

Many plants, shrubs, and trees exude when cut a white, milk-like liquid which is called latex (from Latin "latex" = fluid). Examples are the rubber plants in our houses and the dandelions in meadows and lawns. Some of these latices coagulate on exposure to air to more or less sticky, elastic masses, but air really has not much to do with it. A rubber latex is an emulsion of microscopically small rubber droplets (diameter ca. 0.0005–0.001 mm) in a dilute aqueous solution of proteins, fats, carbohydrates, salts, etc. The rubber droplets have a lower density then the surrounding water solution. But being "lighter", they float to the surface similar to the cream in non-pasteurized milk (milk is also an aqueous emulsion, albeit of protein and fat droplets). The rubber droplets then start sticking together and form a film when the water evaporates.

Several trees of South America and East Asia are especially conspicuous with respect to this behavior. Repeated cutting of their bark enables the gathering of more and more latex. The Maya called such trees "weeping wood" (caa = wood, o-chu = weeping). The name of the trees was then transferred by the European discoverers from the trees to the coagulated mass. The spelling was transformed too: Caa ochu (or quechua) in Maya became cauchuc in old Spanish, and caoutchouc in French.

The two remarkable properties of natural rubber, elasticity and stickiness, were utilized very early. The Maya played with balls from coagulated rubber centuries before Columbus discovered the New World. In the sixteenth century, rubber latex was used as glue to catch birds in East Asia. At about the same time, fabrics were coated with rubber latex in Mexico to give water-proof materials. The latex could not be shipped to Europe, however, since it coagulated during the long journey. Only in 1791, in England, did S. Peal succeed in producing rubber coated fabrics, but this occurred only after L. A. M. Herrisant and P. I. Macquer found turpentine and ether to be solvents for coagulated rubber. Charles Macintosh recognized in 1826 that naphtha, a fraction of petroleum rich in aromatics, was also suitable as solvent for rubber. Even today, the English call raincoats "mackintosh's" or "macs", albeit with a wrong spelling. Fifty years earlier, in 1770, rubber got its English name after E. Nairne discovered its ability to rub away pencil marks much more effectively than bread crumbs. What the English call a "rubber" is named "eraser" in American. Well, England and America are known to be two nations divided by a common language. I know of an Englishman who asked his American secretary rather innocently whether she had a rubber. Whereupon she blushed . . .

The nice mackintoshes had a considerable disadvantage. They were sticky, especially in warm weather. We know today that this tackiness is a direct consequence of the chemical and physical structure of the giant rubber molecules. At that time, however, the thought prevailed that the stickiness of natural rubber was caused by a liquid contained therein. An apparent example for such a liquid containing, sticky material was hide glue (old "carpenter's glue"). This glue was known to yield a thin solution by the addition of plenty of water. Removal of water by evaporation at room temperature or heating at elevated temperatures caused the solution to become stickier. Finally, after all water was gone, the residue became hard and non-tacky.

A similar effect was thought to cause the stickiness of natural rubber. It was only logical to "dry" the rubber by removing the perceived liquid. But how to do that with such a viscous ("tough" in common parlance) material? Heating the rubber converted it into a smelly mass. One could try to enlarge the area of rubber pieces since the rate of evaporation is the higher the larger the surface area. But cut rubber pieces always stuck stubbornly together when they were brought together into intimate contact. Then an Englishman, Th. Hancock, had in 1819 a clever idea: if freshly cut rubber pieces always almost instantaneously stick together, then one has to tear them apart again and again. Hancock made two rolls, each with spikes, which rotated against each other. The rubber between the rolls was constantly torn apart into pieces. The stickiness of the rubber did not disappear through this process, though, since we know now that there was no liquid which could be removed. The rubber became less tough, however, and additives could be worked in far easier. Hancock invented "mastication" (from Greek "mastikhan": to grind the teeth), a process still widely used by the rubber industry.

What else to do to make rubber more useful? An American, Charles Goodyear, felt that a drying agent had to be added to the rubber so that the liquid could be absorbed. Between 1831 and 1839, he systematically tried all drying agents known to him. One of the drying agents tried was sulfur, which even today is used to "dry" wounds (it does not dry them by absorbing the water but it kills bacteria which causes the production of pus to cease). Goodyear also tried to accelerate the "drying" by gentle heating. All attempts were unsuccessful until one day a mixture of rubber, sulfur, and zinc oxide was carelessly left overnight in contact with a hot oven. Low and behold: The stickiness was gone and the previously tacky rubber was converted into a highly elastic, "dry" material. Goodyear recognized the importance of his invention. But he did not find a venture capitalist and thus kept the discovery confidential for five years. It was only 1844 that a patent application was filed. In the meantime, Thomas Hancock got a process for the same process, according to some sources (mainly English) because of an independent invention, according to others (mainly American) because of reverse engineering of a Goodyear sample. Hancock also gave the process its present name. Since heat and sulfur were used and both were attributes of the Roman god Volcanus ("Vulcan"), the process was christened "vulcanization".

This useful invention established the rubber industry. Rubber found more and more applications, and its consumption jumped from 38 tons in 1825 to 388 tons in 1840, and to 8000 tons in 1870, all collected from wild growing trees, mainly in Brazil from trees of the genera *Castilla elastica* and *Castilla ulai*. Several Brazilian rubber barons became very rich and donated handsomely to document their cultural devotion: They built an opera house in the middle of the Amazon jungle, complete with imported Italian marble and Italian singer Caruso.

Brazil practically held the rubber monopoly since the very small East Asian rubber production relied on *Ficus elastica,* the rubber tree of our living rooms. England's maritime interests generated a need of rubber for many applications and it tried therefore to break the Brazilian monopoly. At first, the native *Ficus elastica* was cultivated for plantation purposes in England's East Asian colonies but without success: The latex stopped flowing after several cuttings. The same experiment was repeated with liana rubber from Africa, again without success. Extensive investigations by J. Collins, curator of Physics Gardens of London Apothecaries Company, showed that the genus *Hevea brasiliensis* would be best for plantations. Brazil held however the monopoly for wild rubber and intended to maintain it by embargoing the export of plants and trees. Collins tried in 1873 to smuggle out 2000 seeds but the cargo got shipwrecked. He eventually succeeded in sending 200 seeds to England but only 12 germinated. Six of the young plants were transferred to Calcutta where all died.

Another 70000 seeds were brought to England by the young English coffee farmer Henry Wickham, hiding them under banana leaves. This time, 2800 seeds germinated and 1900 plants were grown in Ceylon. They were the base stock for all plantation rubber, first in Ceylon, then in other East Asian countries. The first 4 tons of plantation rubber were market-ready in 1900; in the same year, wild rubber provided 50000 tons.

Plantation rubber is more economical and homogeneous and less soiled than wild rubber. It has displaced wild rubber almost completely. The present day production of 3800000 tons of natural rubber could not be achieved by collecting wild rubber and the repeated tapping of the wild growing trees of the genus Castilla would lead to enormous environmental damage to the Amazon basin. However, natural rubber still provides only one third of the total world demand for elastomers which is now approximately 13000000 tons.

2.2 Cotton and Its Children

The modern fiber industry developed almost simultaneously with the rubber industry. Fibers were used by man since prehistoric times, either as textile fibers like wool, silk, cotton, and flax, or as hard fibers like hemp, jute, or kapok. Man uses wool the same way as nature does, as insulating material. However, cotton finds an unnatural application as a textile fiber since it is the seed hair of a subtropical bush, and hence does not have the same insulating function as the wool of the sheep or our sparse body hair.

Many attempts have been made to improve cotton for human purposes, in part by tinkering, in part by systematic scientific investigations. Cotton was found to consist of microscopically visible cells. Cotton fibers were thus called "les cellules" by the French botanist Anselm Payen in 1838. Based on Payen's report the French Academy of Science then named the main chemical constituent of cotton "cellulose". This name was probably created from "cellule" and "ose", the latter being the usual final syllable in scientific names for sugars like sucrose or saccharose (cane or beet sugar), lactose (milk sugar), fructose (fruit sugar), maltose (malt sugar), etc. Cellulose is chemically indeed a sugar although it is not sweet, that property requires very small molecules since only those can enter the molecular sized cavities of the chemical receptors of our taste buds and cause the sweet sensation.

Cellulose is a "polymerized" sugar; it is a macromolecule, as we will discuss in detail later. It is the structural component of all plants, including algae, which is a direct consequence of the giant nature of its molecules.

Since cotton was relatively easily grown, attempts were made early to improve its textile properties. John Mercer discovered in 1844 that cotton treated with solutions of caustic soda (sodium hydroxide) leads to fibers with high lustre, increased strength, and improved dyeability. Such "mercerizations" can also be performed with sulfuric acid.

Our modern paper is produced from wood and is a cellulose paper. Since cellulose is the main component of both paper and cotton, it is not surprising (to us!) that L. Figuier demonstrated to the French Academy of Science in 1846 a strengthening of paper similar to that of cotton when it was treated with sulfuric acid. W. E. Gaine received in 1853 an English patent for the same process which delivered a parchment-like material. This artificial parchment does not rot (in contrast to the true parchment from animal hides) and it is also not attacked by insects either.

Six years later, another English patent of Thomas Taylor described how layers of paper sheets can be welded together to give very resistant materials by the combined action of zinc chloride and pressure. The process appeared to be similar to the vulcanization of natural rubber with large amounts of sulfur to hard, black masses (Ebonite or Vulcacite), a process invented in 1851 by Nelson Goodyear, the brother of Charles. Since the rubber and the paper were both "vulcanized" and the latter consisted of fibers, products from the Taylor process were called "vulcanfiber". The material is still used today, for example, for inexpensive suitcases.

We know today that the processes leading to mercerized cotton, artificial parchment, and vulcanized paper fiber are physical conversions of the materials; the chemical nature of the constituent cellulose molecules is not changed. Quite different is the situation if cotton and other cellulose containing materials are not treated by sulfuric acid alone but by a mixture of sulfuric acid and nitric acid. This experiment was unintentionally performed in 1846 by the German Christian Friedrich Schönbein. Chemistry professor at the University of Basle, Switzerland, Schönbein accidentally broke a bottle with a mixture of sulfuric acid and nitric acid in the kitchen of his home and the liquid splashed on the floor. The mixture of these two acids is so aggressive that it was called "aqua regia" (royal water) by the alchemists: It dissolves almost everything, even the otherwise unassailable gold metal. Chemist Schönbein knew this of course and in order to prevent greater harm, he took a piece of fabric to wipe it off. That fabric happened to be the cotton apron of his wife. He rinsed it immediately with water to wash off the acid and dried it near the oven where it promptly inflamed and burnt to ashes. Schönbein had unintentionally converted cotton into gun cotton, i. e., he performed a chemical transformation of cellulose into cellulose nitrate, often also called "nitrocellulose". Cellulose nitrate is quite different from cellulose. It dissolves in many organic liquids in which cellulose is insoluble. A solution in a mixture of alcohol and ether produces, for example, a sticky liquid which can serve to cover small wounds and thus collodium, the first liquid sticking plaster, was invented.

These completely new and unexpected properties inspired other inventors. Cellulose nitrate first led to the discovery of plastics (see below), then to the invention of man-made fibers. The first artificial silk was made in 1879 for the threads of light bulbs by the English scientist Sir Joseph Wilson Swan (1828–1914) who developed incandescent electric light (yes, this was before Thomas Alva Edison (1847–1931)).

At about the same time, a Frenchman, Count Louis Marie Hilaire Bernigaud de Chardonnet (1839–1924) was looking for a replacement for natural silk since the flourishing French silk industry was threatened by the mulberry spinner pest. In 1884, he found a way to spin fibers from cellulose nitrate solutions and to convert the flammable cellulose nitrate fibers into non-flammable cellulose fibers. The new "Chardonnet silk" was exhibited in 1889 at the Paris World Exhibition; two years later, the commercial production started. Many other processes and fibers followed (see Chap. 9).

2.3 The First Plastics

Cellulose nitrate differs from cellulose not only in flammability and solubility but also in another aspect. On gentle heating, it can be molded into hard, elastic objects which is impossible to do with cellulose. Films and coatings can also easily be formed from cellulose nitrate: The first movies were recorded on cellulose nitrate films and thus also called "films".

Alexander Parkes found in 1862 a method for the easy processing of cellulose nitrate after the addition of castor oil, camphor, and dyestuffs. The products were modestly called "Parkesite"; they caused a lot of exitement and even earned a gold medal. A company was founded in 1865 with great hopes and even greater capital investment to produce "Parkesite" but technical difficulties led to bankruptcy two years later. English and American patents of Daniel W. Spill who used alcoholic solutions of camphor instead were equally unsuccessful.

The time was ripe for a technological breakthrough which is usually the result of the ignorance of laymen and the hyperbole of experts. In 1869, John Wesley Hyatt was granted a patent for the use of camphor without castor oil and without alcohol. The mixture of cellulose nitrate and camphor could be shaped at moderate temperatures to hard objects with very smooth surfaces. These were exactly the properties needed for billiard balls, hitherto made from ivory, naturally grown elephant's teeth. Billiard (pool) balls from ivory often have slightly uneven surfaces and they track erratically. Before Hyatt's invention, manufacturers tried to smooth the surfaces by coating them with cellulose nitrate, occasionally with unexpected side effects: Billiard players were literally shooting pool since the surface coating exploded on very hard impacts, which in turn caused some in the Wild West to draw the revolver . . .

Hyatt's invention won the $ 10000 reward for better billiard balls, about $ 100000 in today's money. Many pool balls were however not manufactured from cellulose nitrate/camphor, tradenamed celluloid. This fame is claimed by a later invention, the fully synthetic Bakelite plastic, which is manufactured from phenol and formaldehyde.

The first plastics contributed considerably to wildlife protection since far less elephants were killed for their ivory. More than 8000 elephants died in 1864 just to cover the English consumption of ivory. In 1909 alone, the United States imported more than 400000 kg (880000 lbs) of ivory which was used not only for billiard balls but also for piano keys, knife handles, and haberdashery. This amount corresponds to 9000 elephants since a pair of elephant tusks weighs about 45 kg (100 lbs). It is estimated that world demand for ivory required the annual killing of 100000 elephants during the period 1860–1870.

The new plastics saved annually the lifes of several tens of thousands of elephants. Even people were protected: An estimated 4000 hunters died while hunting these large herbivorous animals. Who thinks of plastics as protectors of the environment? Well, human desires are still at odds with humane feelings: Elephants are continued to be killed for their ivory. The latest statistics discloses that in 1983 Germany imported 32696 kg tusks, 580388 carvings, and 1767 kg of other ivory which translates into at least 2100 animals. The import of Asian ivory (but not of African ivory carvings etc.) is illegal in the United States but what about the use of ivory in other countries such as England, Italy and India, to name a few?

Another early plastic used casein, the white material from skim milk. A "pretty horn" ("schön horn") from skim milk was already described by the German Benedictine monk Wolfgang Seidel (1492–1562); the recipe is said to come from a Swiss merchant, Bartholomäus Schobinger. The formulation was however forgotten until the American Emery Edwin Childs was granted a patent for the production of plastic masses in 1885. In this process, whey is removed from the curds by pressing and the remaining substance is kneaded in hot water until all fats are removed. The mass is finally compacted in molds under the addition of dyestuffs, porcelain powder, etc. This process is purely physical. The chemical reaction of this material with formaldehyde led, however, in 1897 to an artificial horn which the German inventors, Wilhelm Krische and Adolf Spitteler, called "galalith", i.e., "milk stone" (Greek: gala = milk, lithos = stone). Galalith is still used today for haberdashery since it is easy to dye in beautiful colors.

Celluloid, Ebonite and Galalith are semi-synthetic products since they originate from natural products which are then chemically converted into new compounds. The first fully synthetic plastic was invented by the Belgian Leo H. Baekeland who lived in the United States. In 1906, he heated various phenols with formaldehyde and produced insoluble, hard masses. The resulting phenolic resins were resistant against solvents and heat and these "Bakelites" were recognized in 1909 as excellent electrical insulators. They became one of the foundations of the modern electrical industry.

Phenolic resins differ from cellulose nitrate especially in one property, the deformation under heat. Cellulose nitrate can be molded at elevated temperatures to articles which retain the shape after cooling. They can be reshaped many times on renewed heating. Materials with this property are called "thermoplastics" (Greek: "thermos" = warm, "plasticos" = to shape, to form).

Phenolic resins, on the other hand, retain their shape upon renewed heating to the original processing temperature; they cannot be reshaped again by heat. Plastics with this property are called "thermosets". The difference between phenolic resins and cellulose nitrate is due to their different chemical make-up. A short excursion into some basic chemistry is thus necessary before we can proceed with tales of tall molecules. Those who still remember their high school chemistry may of course skip the next chapter.

3 How Big is Big?

... That I might see what secret force
Hides in the world and rules its course.

J. W. von Goethe, Faust I

3.1 Atoms and Molecules

One would expect that such strange phenomena as the stickiness and the vulcanization of natural rubber, the high strength of fibers, or the conversion of cotton into gun cotton would inspire the natural scientists of the nineteenth century to search for their molecular causes. The time was not ripe, however, since such important concepts like "chemical compound", "atom", "molecule", or "chemical bond" were either not created or meant something totally different then today.

The term "atom" (Greek: atomos = indivisible) is based on the philosophical conception of Democritos (about 460-370 B.C.) that all matter must be composed of very small particles which cannot be further divided. The nature of these particles was left open. The word "molecule" (Latin: little mass) was coined in 1811 by Amadeo Avogadro, Conte di Quaregna e Ceretto (1776-1856), for the smallest units of gases. The Frenchman André Marie Ampére (1775-1836) called a "molecule" what we now name an atom and a "particle" what is now termed a molecule. The Swiss botanist Karl Wilhelm von Nägeli (1817-1891) used in 1858 the word "molecule" to characterize the smallest building blocks of starch visible under the microscope; again, not molecules in the modern meaning of the word. After the chemists started to use the word "molecule" to describe a chemical compound containing two or more atoms, Nägeli rechristened his "molecules" and called them "micelles" (Greek: micellos = small crumb). The word "micelle" means today something entirely different, namely the more or less ordered physical aggregation of small molecules to bigger entities without formation of chemical bonds.

Chemists envisaged in the middle of last century that the atoms of a molecule are kept together by "chemical bonds". They soon found that each type of atoms could have only a certain number of bonds. These bonds were later called "valences" (Late Latin: valentia = capacity, strength) and symbolized by lines starting from the symbol of the atom. The hydrogen atom with the symbol H exhibits one valence, the oxygen atom (symbol O) two valences, the nitrogen atom (symbol N) three or four, the carbon atom (symbol C) four, etc.

The immediate consequence of the idea of discrete and constant numbers of valences per atom is the law of constant proportions. A certain molecule always exhibits the same number of atoms: The ratio of the various types of atoms is constant and given by whole numbers. The combination of uni-valent hydrogen atoms H- with di-valent oxygen atoms $-O-$ leads only to water molecules with the composition $H-O-H$ (abbreviated as H_2O) or hydrogen peroxide molecules $H-O-O-H$ (abbreviated H_2O_2) but never to stable molecules with the compositions H_3O or HO_2 (Table 2). The tri-valent nitrogen atom and the uni-valent hydrogen atom can create the ammonia molecule NH_3 or the hydrazine molecule N_2H_4 but not NH_2, NH, N_2H, N_2H_5 etc. There are unstable (short-lived) molecules where the full number of valences is not exercised but these are not called molecules but radicals, cations, or anions (see below).

We have to discuss three other things before we can return to giant molecules: The meaning of the valence bond, the masses of atoms and molecules, and the relationship between molecular structure and properties.

Table 2. Names and formulae of some simple molecules, radicals, cations, and anions

Name	Symbols written with		Abbreviation
	valences	electrons	
Hydrogen atom		H˙	H˙
Hydrogen molecule	H—H	H : H	H_2
Water molecule	H—O—H	H : Ö : H	H_2O
Hydrogen peroxide molecule	H—O—O—H	H : Ö : Ö : H	H_2O_2
Hydroxyl radical	H—O˙	H : Ö ·	HO˙
Hydroxy anion	H—O⊖	H : Ö :	HO⁻
Hydroxylium cation	H — O⊕ — H | H	H : Ö : H H	H_3O^+
Ammonia molecule	H — N — H | H	H : N̈ : H H	NH_3

Table 2 (continued)

Name	Symbols written with		Abbreviation
	valences	electrons	
Ammonium cation	$H-\overset{\displaystyle H}{\underset{\displaystyle H}{\overset{\oplus}{N}}}-H$	$H:\overset{\displaystyle H}{\underset{\displaystyle H}{\overset{\cdot\cdot}{N}}}:H$	NH_4^+
Hydrazine molecule	$\overset{H}{\underset{H}{>}}N-N\overset{H}{\underset{H}{<}}$	$H:\overset{\displaystyle H}{\underset{\cdot\cdot}{\overset{\cdot\cdot}{N}}}:\overset{\displaystyle H}{\underset{\cdot\cdot}{\overset{\cdot\cdot}{N}}}:H$	N_2H_4
Methane molecule	$H-\overset{\displaystyle H}{\underset{\displaystyle H}{C}}-H$	$H:\overset{\displaystyle H}{\underset{\cdot\cdot}{\overset{\cdot\cdot}{C}}}:H$	CH_4
Methyl radical	$H-\overset{\displaystyle H}{\underset{\displaystyle H}{\overset{\cdot}{C}}}$	$H:\overset{\displaystyle H}{\underset{\cdot\cdot}{\overset{\cdot\cdot}{C}}}\cdot$	$CH_3^{\cdot}$
Methyl cation	$H-\overset{\displaystyle H}{\underset{\displaystyle H}{\overset{\oplus}{C}}}$	$H:\overset{\displaystyle H}{\underset{\displaystyle H}{C}}$	CH_3^+
Ethane molecule	$H-\overset{\displaystyle H}{\underset{\displaystyle H}{C}}-\overset{\displaystyle H}{\underset{\displaystyle H}{C}}-H$	$H:\overset{\displaystyle H}{\underset{\cdot\cdot}{\overset{\cdot\cdot}{C}}}:\overset{\displaystyle H}{\underset{\cdot\cdot}{\overset{\cdot\cdot}{C}}}:H$	C_2H_6
Ethylene molecule (ethene)	$H-\overset{\displaystyle H}{C}=\overset{\displaystyle H}{C}-H$	$H:\overset{\displaystyle H}{\overset{\cdot\cdot}{C}}::\overset{\displaystyle H}{\overset{\cdot\cdot}{C}}:H$	C_2H_4

The valence bond was originally introduced as an empirical means to classify molecules. If an atom was given a certain number of valences, then the empirically determined chemical composition of molecules allowed conclusions on how the atoms were linked together. The structure of the molecules could furthermore be related to macroscopically observable

16

properties of molecules (or aggregations thereof), such as volatility, color, melting temperature, chemical reactivity, etc. The reasons for the discrete number of valences per atom and thus for the meaning of a chemical bond remained unclear.

It took many years to find that atoms were not indivisible at all. This discovery led to new insights into the nature of things (praiseworthy unless one is a fundamentalist), the atomic bomb (hardly praiseworthy; yet it seem to have prevented a major war), and nuclear energy (appraisal according to mentality and attitude).

Atoms are composed of at least three different types of elementary particles: protons, neutrons, and electrons. Physicists later discovered a whole

Table 3. Periodic system of chemical elements. Arabic numbers indicate atomic numbers, letters the symbols for the elements. The names of the elements are given in Table A-4 in the Appendix. The numbering of groups follows the recent recommendations of the International Union of Pure and Applied Chemistry

Group

Period	1	2	3	4	5	6	7	8	9	10	11	12	13	14	15	16	17	18
1	1 H																	2 He
2	3 Li	4 Be											5 B	6 C	7 N	8 O	9 F	10 Ne
3	11 Na	12 Mg											13 Al	14 Si	15 P	16 S	17 Cl	18 Ar
4	19 K	20 Ca	21 Sc	22 Ti	23 V	24 Cr	25 Mn	26 Fe	27 Co	28 Ni	29 Cu	30 Zn	31 Ga	32 Ge	33 As	34 Se	35 Br	36 Kr
5	37 Rb	38 Sr	39 Y	40 Zr	41 Nb	42 Mo	43 Tc	44 Ru	45 Rh	46 Pd	47 Ag	48 Cd	49 In	50 Sn	51 Sb	52 Te	53 I	54 Xe
6	55 Cs	56 Ba	57* La	72 Hf	73 Ta	74 W	75 Re	76 Os	77 Ir	78 Pt	79 Au	80 Hg	81 Tl	82 Pb	83 Bi	84 Po	85 At	86 Rn
7	87 Fr	88 Ra	89** Ac	104 Unq	105 Unp	106 Unh	107 Uns											

*Lanthanides 4f	58 Ce	59 Pr	60 Nd	61 Pm	62 Sm	63 Eu	64 Gd	65 Tb	66 Dy	67 Ho	68 Er	69 Tm	70 Yb	71 Lu

**Actinides 5f	90 Th	91 Pa	92 U	93 Np	94 Pu	95 Am	96 Cm	97 Bk	98 Cf	99 Es	100 Fm	101 Md	102 (No)	103 Lw

particle "zoo" but these mesons, quarks, gluons, etc., need not to be discussed for our purposes.

Neutrons are electrically neutral, protons possess a unit positive electrical charge, and electrons a unit negative one. Electrons got their name from amber (Greek: electron) which becomes electrically charged on rubbing against other materials. Protons are always "first" (Greek: protos; see also "prototype" etc.).

All atoms of a given chemical element contain the same number of protons. This number is called the atomic number of the chemical element. Hydrogen has the atomic number 1; it possesses one proton per hydrogen atom. Carbon has the atomic number 6 (six protons), nitrogen the atomic number 7 (seven protons) etc. (Table 3). Since a stable atom must maintain electrical neutrality, each atom of a chemical element must always contain an equal number of (positively charged) protons and (negatively charged) electrons. Helium with the atomic number of 2 therefore possesses 2 protons and 2 electrons.

3.2 Chemical Elements and Compounds

Chemical elements can be arranged in a so-called periodic system in such a way that elements with increasing atomic number but with similar chemical properties like valence numbers or chemical reactivities are on top of each other (Table 3). This arrangement is explained by the older atomic theory as follows:

An atom consists of a dense nucleus of protons and neutrons and a surrounding, far less dense cloud of electrons. In this cloud, electrons are arranged in discrete shells. In the first period of the periodic table, only one shell exists, which can be filled with a maximum of two electrons. In the second period, a second shell with a maximum of eight electrons is added. The third period then has the first shell with 2 electrons, the second shell with 8 electrons and a third shell also with 8 electrons. The fourth and fifth shells contain 16 electrons each.

We need only to look at the electrons in the outer shells of each period in order to understand the chemical structure and properties of giant molecules. Each atom tries to collect the maximum permissible number of electrons for these outer shells. Atoms with this maximum number (2 in period one, 8 each in periods two and three, etc.) are "satisfied" and have no reason to look for partners in other atoms, that is, they remain single-atomic. Elements with such atoms are the noble gases helium (He), neon (Ne), argon (Ar), krypton (Kr), xenon (Xe), and radon (Rn); all are in the eighteenth group of the periodic system (see Table 3). These elements are called "noble gases" because they are gases at room temperature and furthermore so "noble" that they do not react with other elements.

All other elements must seek partners if they want to fill their outer electron shells to the maximum extent. The hydrogen atom with only one electron combines with another hydrogen atom to form a hydrogen molecule $(2 H \rightarrow H:H)$. Each hydrogen atom shares the two electrons (see also Table 2), and hence each atom possesses the maximum permissible number of two electrons for this shell. The molecule is "saturated" and, without further prodding from the outside, is very stable. It is, for example, possible to preserve a mixture of hydrogen molecules and oxygen molecules for a very long time without generating their chemical reaction to water molecules $(2 H_2 + O_2 \rightarrow 2 H_2O)$. If however an electric spark or a catalyst is added, then this mixture will explode violently . . .

The chemical elements of the second and the third period follow the octet rule of G. Lewis (from Latin: octo = eight): Atoms of these elements always try to surround themselves with eight electrons in the outer shell in order to arrive at chemically stable compounds. The chlorine atom lacks only one electron (see Fig. 1) which it can find in another chlorine atom. Each atom in the resulting chlorine molecule is now surrounded by eight electrons

$$: \overset{..}{\underset{..}{Cl}} \cdot \; + \; \cdot \overset{..}{\underset{..}{Cl}} : \quad \longrightarrow \quad : \overset{..}{\underset{..}{Cl}} : \overset{..}{\underset{..}{Cl}} : \tag{3-1}$$

A hydrogen atom may also do as a partner, but now we get a hydrogen chloride molecule, HCl,

$$H \cdot \; + \; \cdot \overset{..}{\underset{..}{Cl}} : \quad \longrightarrow \quad H : \overset{..}{\underset{..}{Cl}} : \tag{3-2}$$

The solution of hydrogen chloride in water is known as hydrochloric acid (unfortunately also symbolized by HCl).

Carbon atoms exhibit four electrons in the outer shell. The combination of a *carbon* atom with four *hydro*gen atoms results in the "hydrocarbon" molecule of methane, the main component of natural gas. If four chlorine atoms join instead, carbon tetrachloride results, formerly used as "dry" cleaning agent "Tetra".

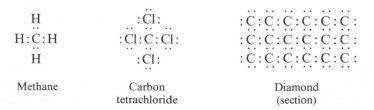

Methane   Carbon tetrachloride   Diamond (section)

And what happens if a carbon atom with four outer electrons joins regularly with four other carbon atoms? The result is a diamond, where every carbon atom unites in a tetrahedral fashion with four others. Since this can continue indefinitely to macroscopic dimensions, each diamond stone is then one single molecule. Some comments about jargon and lingo. Colloquial language is not precise and the jargon of chemists is neither. But chemists (usually) know what they are talking about which cannot always be claimed for journalists. One reads that a dreadful, toxic chemical contains "chlorine", for example in polyvinyl chloride, and immediately sickening green-yellow gas clouds are envisioned. The true "chlorine" of every day is the chlorine molecule Cl_2, indeed a gas, which was used in combat in World War I. But it is not the chlorine atom $Cl^{.}$ and by no means a chlorine which is chemically bound to other atoms. Polyvinyl chloride does contain "chlorine" but as chemically bound substituents and not as absorbed chlorine gas. Chlorine is not "chlorine" if it is chemically bound to other atoms: chlorine (Cl_2) is a green-yellow gas, hydrogen chloride (HCl) a colorless gas, sodium chloride (NaCl; table salt) a white, crystalline material, carbon tetrachloride (CCl_4) a waterclear "heavy" liquid, and polyvinyl chloride $(C_2H_3Cl)_n$ a colorless plastic.

3.3 Valences and Bonds

Table 2 reveals immediately what a valence and a valence bond stand for. A valence is nothing more than an electron pair and a valence bond a symbol for the binding by an electron pair. The electrons of the outermost shell of atoms are correspondingly called valence electrons; only they participate in chemical bonds, at least in the first three periods.

Two atoms may also be bonded by two electron pairs such as the two carbon atoms in ethylene (Table 2). The binding by two electron pairs is usually symbolized by a double line which characterizes the "double" bond. A double bond does not mean a twice as strong bond, however; double bonds are usually much more reactive than single bonds.

Bonds are called covalent bonds if the electrons of the binding electron pair are shared about equally by both bonded atoms. Electrons can be shared in *exactly* equal amounts only when each atom is the same, e.g., carbon atoms surrounded regularly by four other carbon atoms in a diamond.

In all other chemical bonds, one atom contributes more and the other less of an electron to the bond. All chemical elements on the left side of a period in the Periodic Table need only to donate one or a few electrons to a bond in order to reach a fully filled electron shell. All chemical elements on the right side of the Periodic Table readily accept electrons. The binding electrons will not be distributed equally to both atoms: one has more, the other less of the electric charge. Such chemical bonds are called "polar bonds".

So what is the difference to real life? Joining atoms is like joining people in marriage. The bond is stable and both partners have "good chemistry" if both share and share alike. But if one is dominant, then polarization sets in and finally, given the right environment, both partners separate, each one charged up.

Take table salt, scientifically called sodium chloride, NaCl. Both sodium Na and chlorine Cl are farthest away from silicon in the third period (Fig. 1). In a sodium chloride crystal, each atom is surrounded by an electron octet

```
    ..        ..        ..
 Na : Cl : Na : Cl : Na : Cl :
 ..   ..   ..   ..   ..   ..
 : Cl : Na : Cl : Na : Cl : Na
 ..   ..   ..   ..   ..   ..
 Na : Cl : Na : Cl : Na : Cl :
 ..   ..   ..   ..   ..   ..
 : Cl : Na : Cl : Na : Cl : Na
 ..        ..        ..
```

Section of a sodium chloride crystal

The chlorine atoms in a sodium chloride crystal are negatively charged and the sodium atoms positively. Since opposites attract each other and electrical neutrality must prevail, both partners stick together in such a crystal.

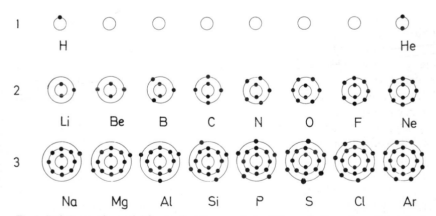

Fig. 1. Bohr's atomic model for the first three periods of the periodic system of chemical elements. Electrons (full dots) circulate around the atomic nucleus (not shown) in fixed orbits. The innermost shell can only accommodate two electrons, the second and third shell offer space for up to eight electrons each. The electrons in the outermost shells are called valence electrons; they are responsible for chemical bonds, for example, four in carbon (C) and six in sulfur (S)

Atoms, like humans, are tempted to escape from their bondage if another attractive partner lures. Water molecules H:Ö:H have non-bonded electron pairs which cause the molecules to adopt an angled structure. The resulting electrical dipole is attracted by an sodium atom. The sodium chloride crystal dissolves in water since the water molecules combine with the sodium atoms which in turn are separated from the chlorine atoms. The sodium chloride "dissociates" into a hydrated sodium cation without "free" outer electrons and a chloride anion with eight electrons in the outer shell:

$$(NaCl)_n \xrightarrow{+H_2O} n\,(Na)^{\oplus}_{H_2O} + n\,(:\overset{..}{\underset{..}{Cl}}:)^{\ominus} \tag{3-3}$$

The sodium cation is positively charged and moves towards the cathode in an electric field. The chloride anion is negatively charged and travels to the anode. Examples of other cations and anions can be found in Table 2.

The alert reader surely noticed that we never talked about a sodium chloride *molecule* and that we also wrote $(NaCl)_n$ in Eq. (3-3), not NaCl. There are indeed no identifiable sodium chloride molecules. NaCl simply does not exist as a pair of atoms (a molecule). It can be misleading to always use the symbol NaCl for sodium chloride, since it can be $(NaCl)_n$ in the crystal, $Na^{\oplus}$ (hydrated) and $Cl^{\ominus}$ in aqueous solutions, or sodium and chlorine atoms in the gas phase. Correspondingly, the formula for the diamond should be written as $(C)_n$.

The ionic bond between sodium (cat)ions $Na^{\oplus}$ and chloride (an)ions $Cl^{\ominus}$ in the sodium chloride crystal is the one extreme of a chemical bond: the third shell electrons surround only the chlorine atoms and the sodium atoms have none. In the other extreme, all outer electrons are shared by all atoms. Examples are the chemical elements sodium, copper, silver, and gold. Since this type of bond is exhibited by metals, it is called a metallic bond. The covalent bond, on the other hand, is in the middle between the ionic and the metallic bond.

Electrons belong to discrete atoms in ionic bonds and are clearly separated from those of other atoms. They also belong to discrete atoms in the covalent bond but are now shared by both partners. In metallic bonds, on the other hand, they are shared by all other atoms and are no longer assigned to individual atoms. Human analogs would be a bitter divorce (ionic bond), a happy marriage (covalent bond), and a commune (metallic bond). Many intermediates exist between these ideal types of chemical bonds since real life is never ideal.

The dissociation of polar bonds into ions is not the only way how chemical bonds can split. On heating molecules with covalent bonds, the atoms may start to vibrate so strongly that the covalent bond is broken and each partner grabs one electron of the bond. Two "radicals" are formed, as in the decomposition of ethane where two methyl radicals are formed

$$H_3C - CH_3 \rightarrow H_3C^{\cdot} + {}^{\cdot}CH_3 \qquad (3\text{-}4)$$

or in the decomposition of hydrogen peroxide into two hydroxyl radicals

$$HO - OH \rightarrow HO^{\cdot} + {}^{\cdot}OH \qquad (3\text{-}5)$$

3.4 Isotopes

Before we tackle the main question of this chapter. "How big is big", we have shortly to return to the structure of atoms. Protons, electrons and neutrons were already identified as the three most important elementary particles of an atom. We also discussed the role of the protons and electrons: The electrons control the chemical reactions and the protons maintain electrical neutrality.

Neutrons, on the other hand, are the silent partners: They exist but hardly influence chemistry of atoms or molecules. They are present in all elements with the exception of the simple hydrogen atom H which possesses only 1 proton and 1 electron. There are however hydrogen atoms which have in addition one or two neutrons, and which are given new names. Deuterium atoms contain one neutron while tritium atoms contain two. Deuterium and tritium are so-called "isotopes" of the simple hydrogen atoms, i. e., atoms with the same number of protons but different numbers of neutrons per nucleus (Greek: isos = equal; topos = place, position).

Only few chemical elements are pure elements with only one type of isotope. An example is fluorine F with the atomic number 9 which accordingly contains 9 protons and 9 electrons but in addition 10 neutrons. Sometimes chemical elements contain small amounts of other isotopes, so to say as contamination or impurity. Hydrogen is an example: It is composed of 99.9844% simple hydrogen (1 proton, 1 electron, 0 neutrons), 0.0156% deuterium (1 proton, 1 electron, 1 neutron) and about 10^{-15}% tritium (1 proton, 1 electron, 2 neutrons). Some chemical elements possess different isotopes in approximately equal amounts whereas others have many isotopes, e. g., tin with 10 isotopes.

3.5 Masses and Molar Masses

Protons, neutrons, and electrons are very small entities (Table 4). A proton weighs only about $1.67 \cdot 10^{-25}$ g, that is

0.000 000 000 000 000 000 000 000 167 g

Name	Mass in g (rounded)	Molar mass in g/mol (rounded)	Unit charge
Neutron	$1.675 \cdot 10^{-24}$	1.008 77	0
Proton	$1.673 \cdot 10^{-24}$	1.007 57	$+1$
Electron	$9.110 \cdot 10^{-28}$	0.000 55	-1

or

0.000 000 000 000 000 000 000 000 369 lbs

Such small masses are inconvenient. In chemical calculations, either relative masses or molar masses are thus used instead of the masses themselves.

A relative molecular mass M_r is defined as the ratio of the mass m of an elementary particle, atom, or molecule to the mass of an atom of the carbon isotope 12 (6 protons, 6 neutrons, 6 electrons)

$$M_r = m/m_{12} \tag{3-6}$$

The relative mass of an atom was formerly called "atomic weight", the relative mass of a molecule "molecular weight".

Macroscopic amounts of chemical elements or compounds contain many, many molecules, for example, 12 grams of the carbon isotope 12 contain about $6.02 \cdot 10^{23}$ carbon atoms. It is thus expedient to define this number as the basic unit of chemical elementary particles such as molecules, atoms, protons, electrons, etc., and to assign it a special physical unit, the "mole" (abbreviation: mol). The so-called Avogadro number N_A gives the number of particles per mole particle

$$N_A = 6.022\ 52 \cdot 10^{23} \text{ particles/mole particle} = 6.022\ 52 \cdot 10^{23} \text{ mol}^{-1}$$

Molar masses are defined as products of the corresponding absolute masses with the Avogadro number

$$M = m \cdot N_A \tag{3-7}$$

Relative masses and molar masses are based on the same source, they also possess the same numerical value provided that the molar mass is measured in g/mol. Only the physical units (formerly called "dimensions") are

Table 5. Molar masses, lengths of the stretched chains, and diameters of various macromolecules

Macromolecule	Molar mass in g/mol	Length in mm	Diameter in nm
Deoxyribonucleic acids from			
Lung fish	69 000 000 000 000	34 700	2.0
Escherichia coli bacteria	2 500 000 000	1.3	2.0
Polyoma SV 40 virus	3 000 000	0.0015	2.0
Starch			
Amylopectin of Easter Lily	250 000 000	129	0.74
Amylose of wheat starch	2 000 000	1.0	0.74
Polyethylene			
ultra-high molar mass	3 000 000	0.027	0.49
conventional	100 000	0.00091	0.49

different: Relative masses have the physical unit of unity (they are "dimensionless"), molar masses however the unit "mass per amount of substance" (for example, g/mol).

How big are giant molecules or macromolecules (Greek: makros = big)? That is a matter of view. As chemistry became an exact science during the last century, chemists first investigated the readily accessible and easy to investigate small molecular compounds. These compounds have low molar masses: Water with 18.015 g/mol, hydrazine with 32.045 g/mol, methane with 16.043 g/mol, ethyl alcohol with 46.069 g/mol, cane sugar with 342.30 g/mol, etc. Measured by their molar masses, macromolecules are indeed giant: Polyethylenes for packaging films have molar masses of about 100 000 g/mol, those for artificial knee joints about 3 000 000 g/mol (Table 5). The world record for single, isolationable molecules is presently held by the deoxyribonucleic acids (DNA) of the lung fish which have molar masses of about 69 million millions (69 American trillions or 69 English billions) g/mol, truly giant molecules. Giant however only relative to other molecules since the absolute mass m of even this giant is very small:

$$m = \frac{M}{N_A} = \frac{69 \cdot 10^{12}\,\text{g} \cdot \text{mol}^{-1}}{6.02 \cdot 10^{23}\,\text{mol}^{-1}} = 1.15 \cdot 10^{-10}\,\text{g} = 115\,\text{pg} \tag{3-8}$$

One can calculate that such a molecule, when stretched, must be very long, 34,7 m in our example (1138 feet). Yet it is very thin, only $2 \cdot 10^{-9}$ m = 2 nm. The resolving power of the human eye is too low in order to see such structures. Even the normal (optical) microscopes cannot do it but electron microscopes with a resolution of ca. 0.3 nm make such molecules visible (see Chap. 7).

25

4 False Doctrines

> *. . . yet the joy is unsurpassed*
> *of insight into eras long ago,*
> *to see how wise men then thought thus and so*
> *and how we reached our splendid heights at last.*
>
> J. W. von Goethe, Faust I

4.1 The Discovery of Polymerizations

Why did chemists not find out that big molecules composed of thousands or even millions of atoms exist besides "normal", small molecules of 2–50 atoms, even after the modern definition of a molecule was introduced? Well, there were many prophets but they only had to offer beliefs, not convincing experiments. Too many observations seemed to contradict the existence of macromolecules.

Experience showed that the postulated chemical bonds between atoms were very strong, whatever their nature was. The chemical structure of molecules typically stays constant over a wide temperature range, even if their physical state changes drastically. Molecules with the composition H_2O form liquid water at room temperature. At temperatures below $0\,°C$ ($32°\,F$), liquid water is transformed into solid ice, at temperatures above $100\,°C$ ($212°\,F$) and at normal atmospheric pressure into gaseous water vapor. Liquid water reappears on cooling the water vapor or on melting the ice. Practically all chemical compounds investigated at that time showed the same sequence of physical states with increasing temperature:

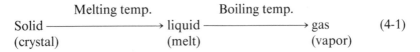

$$\text{Solid} \xrightarrow{\text{Melting temp.}} \text{liquid} \xrightarrow{\text{Boiling temp.}} \text{gas} \qquad (4\text{-}1)$$
$$\text{(crystal)} \qquad\qquad \text{(melt)} \qquad\qquad \text{(vapor)}$$

There were however unexplained exceptions. The tree *Liquambar orientalis* from Asia minor delivers an aromatic resin called "storax" which was used over 3000 years ago by the Egyptians to embalm their dead. Storax was traded in many grades, for example, as "liquid storax" *(styrax liquidus)*, a dark grey, highly viscous mass.

On heating liquid storax with water, Eduard Simon observed in 1839 that a clear organic liquid distills off which he named "styrol" ("styrene" in English). Simon found that further heating of styrene to 200 °C resulted in a solid gelatinous mass. This transformation of a liquid into a solid upon heating was remarkable since liquids normally change to gases (and not to solids) above a certain temperature. Simon suspected a chemical reaction of the styrene, and since the heating was done in air and oxygen was the only possible reaction partner present in the air, he postulated a reaction of the styrene with the oxygen (an "oxidation") and called the gelatinous product "styrene oxide".

The product was analyzed six years later by John Blyth and August Wilhelm (von) Hofmann (1818–1892, knighted in 1888). These researchers found no oxygen in the alleged styrene oxide and changed its name to "metastyrene". The name was not explained in their paper, but it seemed to be very obvious to the learned chemists of the times: metastyrene is found "after" styrene ("meta" is the Greek word for "after" or "under", "between", "besides", "above", etc.).

Most remarkable was however that metastyrene had the same composition (in the old sense as ratio of atoms) as styrene itself, i. e., eight carbon atoms and eight hydrogen atoms. Styrene and metastyrene thus had the same composition but different properties. Such chemical compounds had been called "isomers" by the Swedish chemist Jöns Jacob Berzelius (1779–1848) in 1833 (Greek: iso = equal, meros = particle). A subgroup of these isomers, those with equal ratios of atoms but different (i. e., many) molecular sizes, was named "polymers" by Berzelius (Greek: poly = many). Examples given by Berzelius for such polymers were ethylene (C_2H_4) and butylene ($C_4H_8 = 2 \cdot$ "C_2H_4")

Isomers

$CH_2=CH_2$	$CH_2=CH-CH_2-CH_3$	$CH_2=C\begin{smallmatrix} \diagup CH_3 \\ \diagdown CH_3 \end{smallmatrix}$
Ethylene	Butylene (butene-1)	Isobutylene (isobutene)

Polymers

According to modern nomenclature, butylene and isobutylene are indeed isomers but ethylene and butylene are not. The reason is the different meaning of the word "composition" which in the times of Berzelius indicated the *ratio* of the different types of atoms (always 1 carbon to 2 hydrogen in the examples). Modern nomenclature emphasizes the absolute *number* of atoms per molecule. Butylene and isobutylene have the same composition (C_4H_8) whereas that of ethylene (C_2H_4) is different.

Blyth and Hofmann did not use the words "isomer" and "polymer" but these terms must have slowly become part of the general vocabulary of chemists since the French scientist Marcelin Berthelot (1827–1907) described in 1866 in his investigations on "isomers" of acetylene that styrene changes on heating to a "polymère resineux" (resinous polymer). Styrene and metastyrene were assigned the same composition $C/H = 16/8$ (instead of the proper $C/H = 8/8$) since Berthelot, in accordance with many other researchers of that time, assigned carbon the relative atomic mass of 6 instead of 12. Berthelot called metastyrene a polymer since it was a "polymer" of acetylene which he assumed as $C/H = 4/2$ (instead of $2/2$) and not because it was a polymer of styrene (which, according to this old nomenclature, is also a polymer of acetylene). Since the "polymer" metastyrene was formed on heating styrene, the process was obviously a "polymerization", the first time that this word was used.

"Polymerization" thus originally meant that a number of molecules were united to a bigger entity without change of the ratio of atoms. Styrene was not the only chemical compound which at that time was known to undergo such a polymerization. The French chemist Adolf Wurtz (1817–1884) discovered the polymerization of the gaseous ethylene oxide (C_2H_4O) to liquids with the "same" composition but various properties. At the same time, A.-V. Lourenço observed that the reaction products of ethylene oxide showed the less deviation from the formula C_2H_4O the higher the viscosity of the polymerization products was. He also assigned the liquid reaction products a chain formula, in modern symbols

$$n \, H_2C\!-\!\!\!-\!\!\!-\!CH_2 \rightarrow -(CH_2-CH_2-O)_n-\, ; n = 1-6 \qquad (4\text{-}2)$$
$$\diagdown\diagup$$
$$O$$

Gaseous   Liquid
ethylene oxide   polyethylene oxides

The nature of the atoms at the ends of the chains was left open, as indicated in Eq. (4-2) by dashes.

A similar chain formula was used in 1869 by A. Kraut to describe the "polyester" generated from salicylic acid acetyl ester; in the symbolism of the times

$$H \, . \, O \, . \, C_6H_4 \, . \, CO \, . \, O : 6(C_6H_4 \, . \, CO \, . \, O)C_6H_4 \, . \, CO \, . \, O. \, H$$

Today, we would write

$$H - O - C_6H_4 - CO - O - (C_6H_4 - CO - O)_6 - C_6H_4 - CO - O - H$$

The old and the new formula are really not very different. Changes were made in the interpretation of the dots (old) and dashes (new). We identify today the single dashes with an electron pair which binds two atoms chemically (see Chap. 3). Oh, by the way, "polyesters" are a *class* of polymers, not a single compound. Kraut's polyester is quite different from the polyester used today in textiles (see Chap. 9).

4.2 Carbon Chains

Chemists are always very ingenious to depict facts graphically. Polymethylene, for example, is composed of many methylene groups (CH_2) which are lined up like pearls on a string or chain. The resulting polymethylene "chain" $-(CH_2)_n-$ can be represented in many different ways (Fig. 2), depending on whether one wants to emphasize the type of the chemical bonds, the angle between chain bonds (valence angle), the bond lengths, or the space filled by the atoms. If one is mainly interested in the chain bonds, then a formula with electron pairs or valence bonds or the corresponding short-hand notations may be useful. A wave line at the two ends of the chain may indicate that the chain extends further. One normally need not worry about the chemical character of the chemical groups at the ends of chains (hence the wave line) since these "end groups" do not influence most properties of macromolecular compounds. In many cases, the nature of endgroups of macromolecular chains is not determined.

The short-hand notations of chain formulae are especially useful if one is just interested in the chain character of a macromolecule and not in the chemical bonds. An even shorter short-hand notation is sufficient if only the constitutional repeating unit is to be demonstrated, in our example the methylene group $-CH_2-$.

A carbon atom possesses four valences which are all equivalent in true covalent bonds. The valences thus must be distributed equally in space (Fig. 3a). The four chemical bonds between the carbon atom and the four hydrogen atoms in a methane molecule CH_4 thus correspond to the four lines radiating from the center of a tetrahedral pyramid (a tetrahedron; from Greek: tetraedros = four-faced) (Fig. 3b). The angles between each pair of bonds are identical and constant, and according to the general laws of geometry must be 109 degrees and 28 minutes. These angles are called the "valence angles" of the carbon atom.

What happens if we replace two of the four hydrogen atoms of a methane molecule with two methylene groups? You guessed right, we now have a section of a polymethylene chain. By doing so, valence angles of the carbon atoms are only little changed. They do not stay constant, however,

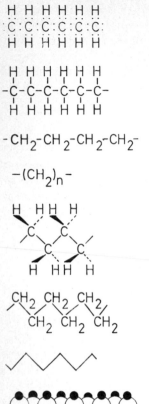

H H H H H H
:C:C:C:C:C:C:
H H H H H H

Formula with electron pairs

H H H H H H
-C-C-C-C-C-C-
H H H H H H

Formula with valence bonds

$-CH_2-CH_2-CH_2-CH_2-$

Short-hand "chain" formula with valence bonds

$-(CH_2)_n-$

Short-hand formula

"Three-dimensional" formula; atoms on wedges project out of the paper plane, atoms on broken lines into the paper plane

CH_2 CH_2 CH_2
CH_2 CH_2 CH_2

Short-hand of the three-dimensional formula

Abstracted three-dimensional formula

Projection of a space-filling model into the paper plane

Fig. 2. Some often used representations for the chemical structure of polymethylene $R-(CH_2)_n-R$. The endgroups R are left out. Double points, dashes, wave lines, etc., are used to indicate that the chain structure extends "infinitely"

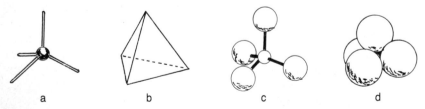

a   b   c   d

Fig. 3. Molecular models of the carbon atom. (a) shows the carbon atom as black ball with four radiating valences, (b) the resulting tetrahedron if the corners of the valences are connected. (c) depicts a carbon atom with four equal substituents, as in e. g. methane CH_4; the space filling properties are indicated by spheres. (d) shows the same molecule as in (c) albeit with the correct relative sizes of spheres since hydrogen and carbon atoms have different diameters and thus different space requirements. In addition, $C-C$ and $C-H$ bonds differ slightly in polarity. Both effects cause the valence angle to change from $109°28'$ in $H-C-H$ of CH_4 to ca. $112°$ in $C-C-C$ of $-(CH_2)_n-$.

A zig-zag chain results if all methylene groups are lined up regularly (Fig. 2). If the carbon atoms of a polymethylene chain are placed in the paper plane of Fig. 2, then some hydrogen atoms must stick out of the paper plane and some must protrude below the plane since the valences of the carbon atoms are distributed tetrahedrally (Fig. 3).

The space filled by atoms may be represented by spheres. However, only rarely do chemists use the true relative sizes of atoms, since hydrogen atoms are sometimes too large relative to carbon atoms. See Fig. 3 d where the carbon atom is buried under the hydrogen atoms and valence bonds can no longer be seen. The atoms are thus often made smaller than they really are (Fig. 3 c) just to allow a look at the valence bonds. Sometimes the right relative size is shown as in the bottom of Fig. 2.

4.3 What is a Polymer?

A chain formula seems to be self-evident: if atoms possess at least two valences, then they may be coupled to each other like railroad cars in a train. The railroad cars may, however, be firmly connected to each other or they may just touch each other as it is done on switching. This was exactly the problem chemists faced during the last century: Were the strange properties of polymers caused by very big molecules with covalent bonds or were they just assemblies of small molecules held together by weaker physical (non-covalent) bonds?

The question is by no means trivial. Scientists knew of the existence of weak "physical" bonds besides true covalent bonds. These physical bonds were called "secondary valences" "for just where no ideas are, the proper word is never far" (J. W. von Goethe, Faust I). An example for the existence of such secondary valences is water, H_2O, the chemical compound from one oxygen atom with two hydrogen atoms. All chemical elements of the Periodic Table surrounding oxygen form with hydrogen compounds which are gases at room temperature: Nitrogen, to the left of oxygen, gives ammonia (NH_3), fluorine, to the right of oxygen, forms hydrogen fluoride ("HF"), and sulfur, below oxygen, gives hydrogen sulfide (H_2S) (see Table 3). Why, then, is water a liquid if all neighbors are gases? Because of the secondary valences, that is why . . .

Thomas Graham made, in 1861, another curious observation. He noticed that solutes of aqueous solutions travelled with different speeds through membranes: those from crystalline solids much faster than those from non-crystalline solids. Members of the second group were called "colloids" by Graham since bone glue was an especially prominent representative (Greek: kolla = glue). It was found later that even crystallizing compounds may give "colloidal", i.e., highly viscous, solutions with slowly diffusing solutes. The chemical reactivity of low viscous "crystalloid" solutions and high viscous "colloid" solutions was however not very different. It was thus concluded that colloids were nothing but physical assemblies of small molecules held together by secondary bonds. They can not be big molecules, so went the reasoning, since covalent bonds change the chemical reactivity and the high viscosity colloidal compounds should thus differ in their chemical reactivity from that of their low viscosity, crystalloid homologs, which they did not.

The problem – big real molecules vs. assemblies of small molecules – can be decided by molar mass determinations. But scientists did not have the required methods during the first eighty years of the nineteenth century. All methods known to chemists and physicists were chemical methods, using the composition of compounds. These methods failed, however, since compounds with such different properties as the liquid styrene and the solid (yet non-crystalline) metastyrene had the same relative chemical composition.

François-Marie Raoult (1830–1901) discovered in 1882 and 1883 that a relationship exists between the vapor pressure of the solvent above a solution and the molar fraction of the solute in the solution. Jacobus Hendricus van't Hoff (1852–1911) established in 1887/1888 a theoretical law which connected osmotic pressure, concentration, temperature and molar mass. The law allowed the determination of molar masses from physical properties without requiring knowledge of the chemical composition of the molecules.

However, that made the Raoult and van't Hoff laws suspect to many chemists. Was it not strange that these new, physical methods gave very high molar masses of up to 100000 g/mol for colloids like rubber, cellulose nitrate, starch, and egg albumin (Table 6), something which was never found for crystalloids? Was it not suspicious that not only different researchers but sometimes also the same scientist found various molar masses for the "same" chemical compound? Furthermore, why did the calculated molar masses often change with the concentration of the colloids whereas those of crystalloids stayed constant if chemical methods were used? Was this not a violation of the laws of Raoult and van't Hoff?

Was it not also the truth, and nothing but the truth, that pure chemical compounds always crystallize but impure ones did not? This truth was based on many experiences, for example, with cane sugar. The non-crys-

Table 6. Some early determinations of molar masses of natural macromolecules

Chemical compound	Molar mass in g/mol	Method	Year	Author(s)
Natural rubber	6 000–12 000	Osmotic pressure	1888	Gladstone and Hibbert
Natural rubber	100 000	Osmotic pressure	1914	Caspari
Hemoglobin	16 700	Chemical analysis[a]	1886	Zinoffsky
Hemoglobin	68 000	Ultracentrifugation	1925	Svedberg
Hemoglobin	66 700	Osmotic pressure	1925	Adair
Egg albumin	14 000	Depression of freezing temperatures	1891	Sabanjoff and Alexander
Egg albumin	17 000	Diffusion	1908	Herzog
Egg albumin	73 000	Diffusion	1910	Herzog
Egg albumin	34 000	Osmotic pressure	1919	Sørensen
Egg albumin	43 000	Osmotic pressure	1925	Sørensen
Egg albumin	45 000	Ultracentrifugation	1925	Svedberg and Nichols

[a] Hemoglobin consists of four approximately equal "subunits"; the analysis of Zinoffsky refers to the subunit.

Found by Zinoffsky

for the su unit $C_{712}H_{1130}O_{245}N_{214}S_2Fe$ $(M_r = 16\,728)$

ie., for hemoglobin $C_{2848}H_{4520}O_{980}N_{856}S_8Fe_4$ $(M_r = 66\,912)$

Today's value

for human hemoglobin $C_{3108}H_{4602}O_{890}N_{742}S_{12}Fe_4$ $(M_r = 67\,209)$.

tallizable colloids simple could not be pure compounds and hence no self-respecting chemist would touch such stuff.

Not all chemists could afford to be academic purists, however. Some were forced by their jobs to tackle these strange substances. One was Samuel Shrowder Pickles at the Imperial Institute London. Great Britain still had a vital interest in natural rubber whose chemical constitution was not fully explored. Pickles, in 1910, suggested for natural rubber a chain structure, in modern symbols

$$\sim CH_2 - \underset{\underset{CH_3}{|}}{C} = CH - CH_2 - CH_2 - \underset{\underset{CH_3}{|}}{C} = CH - CH_2 - CH_2 - \underset{\underset{CH_3}{|}}{C} = CH - CH_2 \sim$$

Earlier oxidation experiments by the German Carl Harries did not indicate any endgroups. Harries thus assumed natural rubber to consist of small rings with two isoprene units $-CH_2 - C(CH_3) = CH - CH_2 -$ each. Pickles also concluded that the head of the chain must bite its tail: The rubber molecules were envisaged as closed "rings" (= cyclic molecules) or closed chains with more than two isoprene units and not as open chains.

33

The apparent absence of endgroups can, of course, also be explained by the low sensitivity of the experimental methods towards a small amount of endgroups in very long open chains. No researcher seriously considered this possibility, however. Why not? Because chemists were mainly thinking qualitatively, not quantitatively, in those years.

Pickles furthermore argued that physical and chemical bonds should be distinguished from each other by chemically modifying the rubber. Harries had assumed that rubber molecules consist of rings of only two isoprene units. The colloidal character of rubber solutions was thought to arise from the association of many such molecules held together by intermolecular "secondary valences". Pickles maintained that a chemical modification of the rubber molecules must modify these secondary forces and should thus depend on the environment of the molecule. Either a change of the solvent or the change of the chemical make-up of the rubber should thus alter the secondary valences.

Pickles thus added bromine (Br_2) to the double bonds of the rubber molecules. The resulting rubber bromide

$$\sim CH_2-CBr-CHBr-CH_2-CH_2-CBr-CHBr-CH_2-CH_2-CBr-CHBr-CH_2\sim$$
$$\qquad\quad | \qquad\qquad\qquad\qquad\quad | \qquad\qquad\qquad\qquad\quad |$$
$$\qquad\quad CH_3 \qquad\qquad\qquad\qquad\quad CH_3 \qquad\qquad\qquad\qquad\quad CH_3$$

likewise exhibited colloidal properties. Since rubber and rubber bromide were unlikely to have the same secondary valences, Pickles argued that both compounds could not be associations of small molecules but must be genuine "large" molecules held together by covalent bonds.

The establishment was not impressed and Pickles' paper was not recognized. There were simply too many disproofs of the molecule theory of organic colloids: equal composition and equal chemical reactivity of colloids and their monomers, often easy chemical transformations from the monomer to the colloid and back, anomalies in the molar mass determinations by physical methods, and the impossibility to crystallize these colloids. The apparent absence of endgroups was taken as ultimate proof for the presence of "micelles" of small rings.

Enter Hermann Staudinger (1881–1965). This German chemistry professor was at that time teaching at the Swiss Federal Institute of Technology in Zürich (ETH). Staudinger had investigated ketenes, a group of low molar mass carbon compounds, for many years and he believed that the basic structure of these molecules was a ring of four carbon atoms (a cyclobutane ring). Another author pleaded in favor of dimers from two molecules held together by secondary valences. Staudinger, in self-defense, collected all pro's and con's for true covalent bonds in these molecules and, by extension, also for organic colloids. In many years with many coworkers, he tried to disprove the so-called micelle theory of organic colloids. It is amusing to see that macromolecular chemistry owes its life to the defense of the structure of small molecules.

The long march through the experimental findings was full of obstacles which Staudinger tried to remove with more and more experiments. Like Pickles, he employed chemical transformations of the organic colloids, curiously without quoting the former although Staudinger referred to the Pickles paper in another publication. Pickles compared the colloid character of the molecules before and after the transformation however only qualitatively whereas Staudinger used a quantitative method: the number of constitutional repeating units per molecule. This number was the same before and after the transformation, strong evidence for covalently bound molecules and against micelles from small molecules. Staudinger thus postulated the presence of large molecule chains, macromolecules (Greek: macros = big).

Such open chains should possess endgroups but these groups were not found. Possible explanations were

a) no endgroups were present since the molecules are rings, not open chains,
b) the chemical reactivity of the endgroups decreased with increasing chain length so that none could be found by chemical methods, or
c) the molecules were so large and the analytical methods so insensitive that the endgroups remained undetected.

Most chemists bet on a), whereas Staudinger favored b). All were wrong: c) was true. But who is really admitting that his (almost no "her's" at that time!) experimental methods are not good enough . . .

Point b) does indeed contain a tiny grain of truth. The chemical reactivity of a molecule is determined by two factors: the reactivity of a group and the number of such groups per molecule. Suppose a so-called carboxylic group COOH has the same reactivity per *group* in acetic acid $H-CH_2-COOH$ and in $H-(CH_2)_{1000}-COOH$. The reactivity per *molecule* must be smaller by a factor of 1000 for the latter, since there are thousand times more methylene groups per molecule and the COOH group is thus diluted by the methylene groups for reactions with other partners.

The principle of equal chemical reactivity of groups was explicitly recognized in the thirties by Paul J. Flory (1910–1985). This scientist thus removed an important mental obstacle for the synthesis of macromolecules. If the reactivity per group is indeed decreasing with increasing molecule size, as assumed by many others, then no macromolecules could be formed at all since the reactivity would be zero at infinite lengths. Who would then do doctoral theses ...

The X-ray measurements of partially crystalline polymers by the I.G. Farben scientists Kurt H. Meyer (a German, later emigrated to Switzerland) and Hermann F. Mark (an Austrian, later emigrated to Canada and the U.S.) also confirmed the basic correctness of the Staudinger hypothesis of molecule chains. These authors did not agree, however, with Staudinger's "very high" molar masses of some ten thousand. Their X-ray measurements rather indicated small crystallite lengths of only 30-60 nm. Assuming that molecules can not be longer than a crystallite (and all data of that time pointed towards that rule without exception), a molecule could have molar masses of no more than 5000-10000 g/mol. We know today that a crystallite can be much shorter than a molecule since chain-like macromolecules can fold upon themselves.

4.4 Necessity Fathers Inventions

Chain folding was discovered in 1957 and was thus unknown to Staudinger in the twenties. Staudinger had to look for other means to prove that his "high" molar masses were indeed correct. He chose viscosity measurements albeit out of desperation since it was an unproven method. Staudinger would have prefered to use a so-called analytical ultracentrifuge, but the authorities were not willing to spend that much money for this newly invented and very expensive instrument. Sounds familiar, doesn't it?

Why an ultracentrifuge? Staudinger was mainly working with synthetic macromolecules which he nonetheless considered models for biological polymers. He thus always followed the literature on the newest developments in colloidal natural compounds. His love of biopolymers was not reciprocated, however, since the "true" biochemists did not pay any attention to the quibbles about the nature of synthetic organic colloids amongst the organic and physical chemists. Whether this was because of the "special nature" of biological compounds, a still existing variety of the *"vis vitae"*, or whether one considered Staudinger's ideas as too farfetched – we do not know.

Fact is however that the idea of macromolecules was developed in biochemistry practically independent of the same idea in organic and physical chemistry. James B. Sumner (1887-1955) succeeded in 1926 to crystallize the enzyme urease. Enzymes form colloidal aqueous solutions. Colloids could thus be crystallized without loosing their chemical reactivity. The organic chemists did not pay much attention, however ...

The Swedish professor Theodor Svedberg (1884-1971) showed in 1924-1927 with his newly invented ultracentrifuge that an enzyme always exhibited the same molar mass, regardless of the conditions. These ultracentrifuges subject molecules and particles to gravity forces up to 500 000 times the earth's gravity (500 000 G). For comparison: A jack-rabbit start of a car pushes you into your seat with 1.5 G; astronauts have to survive "only" 9 G. The high gravitational forces in the rotor of an ultracentrifuge causes molecules to move, similar as sand particles move towards the bottom of a lake under the influence of the earth's gravity. Bigger molecules move faster than small ones and Svedberg could thus calculate the molar masses of enzymes from their sedimentation rates during ultracentrifugation.

Constant molar masses of an enzyme under various conditions (types and concentration of salt solutions, temperatures) meant that the observed high molar masses of enzymes like hemoglobin (see Table 6) could not be caused by secondary forces between smaller subunits. Enzymes thus had to be macromolecules. Arne Tiselius (1902-1971), also a Swede, found later, in 1938, that enzymes always carried the same electrical charge per mass, quite different from the behavior of inorganic colloids. This finding also backed the idea of the macromolecular character of enzymes.

Since Staudinger did not get his expensive ultracentrifuge and thus could not use this instrument to study the molar masses of synthetic organic macromolecules, he had to resort to the far-less expensive viscometry, a very simple experimental method. The theory of viscometry is however far more complicated than Staudinger suspected:

Lourenço observed already (Chap. 4.1) that the viscosity of polyethylene oxides, $HO(CH_2CH_2O)_nH$, increases with increasing molecule size, i.e., increasing degree of polymerization n. The viscosity of melts does indeed allow conclusions on the molar mass of their constituent macromolecules. The melt viscosities of the compounds investigated by Staudinger were however so high that they could not be measured by him with his means. He did not want to do this anyway since he was interested in the molar mass of single molecules and not in the complex interactions which occur in assemblies of molecules such as melts. Honey and water glass are two examples where low molar mass compounds interact so strongly in concentrated aqueous solutions that the contributions of the molar mass to the observed viscosity is negligible compared to the contributions of the intermolecular interactions. Honey is a very concentrated solution of various sugars in water. Water-glass is an aqueous solution of "poly"silicic acid, $\sim\sim (Si(OH)_2 - O)_n \sim\sim$, with n = 4, 8, etc., but certainly not n = 1000.

In order to get the contribution of a single molecule to the observed solution viscosity, one has obviously to work in very dilute solution and not with melts. The more dilute, the less molecules are present per unit volume and the less interactions are possible. The viscosity of such very dilute solutions is then given by the contributions of the solute and the solvent. The difference between the viscosity of the solution, η, and the viscosity of the solvent, η_o, is thus a measure for the viscosity of single molecules of the solute. This difference then has to be related to the viscosity of the solvent to give the so-called specific viscosity $(\eta - \eta_o)/\eta_o = \eta_{sp}$. The viscosity furthermore increases with the concentration c of the solute; division of the specific viscosity by the concentration results in the reduced viscosity, $\eta_{red} = \eta_{sp}/c$. Remaining effects of interactions are eliminated by plotting reduced viscosities against concentrations and extrapolation to zero concentration. The resulting "intrinsic viscosity" $[\eta]$ can be a measure of the size (and shape) of the solute molecules.

But how does one relate the intrinsic viscosity to the molecular size of the solute? Albert Einstein (1879–1955) deduced theoretically in 1905–1910 that the intrinsic viscosity of hard spheres is always $2.5/\rho$, regardless of the sphere size (ρ = density). The molar mass of sphere-like molecules can thus not be calculated from viscosity measurements.

Staudinger assumed that chain-like macromolecules form very rigid chains such as in Fig. 2. A rigid chain looks from the outside like a rod or cylinder. The extension of such a rod by one repeating unit must then increase the observed intrinsic viscosity. This increase was (erroneously) assumed by Staudinger to be directly proportional to the molar mass of a repeating unit

$$[\eta] = K \cdot n \cdot M_R = K \cdot M \tag{4-3}$$

The molar mass M of a molecule is given by the molar mass of the repeating unit, M_R, multiplied with the number n of such units, i.e., the degree of polymerization. K is a proportionality constant which varies with the system solute/solvent/temperature but is independent of the number or repeating units per molecule.

Viscosity measurements were the fairy wand which Staudinger and his disciples used to investigate many compounds. Nothing even changed after theoreticians rightfully insisted that viscometry is at best a kind of black magic since it is not as simple as Professor Staudinger thought.

W. Haller showed already in 1931 that chains are by no means that rigid as Staudinger assumed. Chain atoms can rather rotate around their valence bonds as we will see in detail in Chap. 7. The Swiss Werner Kuhn (1899–1963) concluded from this rotation around chain bonds that macromolecules must form coils, not rods. A macroscopic example for such a behavior is a garden hose. A short hose is rigid and rod-like. Long hoses are not straight, however, they bend in all directions and form a coil.

The Staudinger school did not want to be deprived of their idea of rigid rods and they insisted even in 1941, at a party to honor Staudinger's 60th birthday: "... Kuhn's coils are an abomination to us ..." ("... die Kuhn'schen Knäuel sind uns hier ein Greuel ..."). Yet Lars Onsager (1903–1976) showed already in 1932, before Kuhn, that long ellipsoids must have much higher intrinsic viscosities than those found experimentally by H. Staudinger. We know today that the intrinsic viscosity of rods without rotational diffusion depends on the square of the molar mass whereas those of coils with no net solvent interactions show square root dependencies.

The idea of chain-like chemical structures of compounds like natural rubber, cellulose, polystyrene, and polyvinyl chloride slowly took hold during the early thirties. The existence of macromolecules was no longer doubted after World War II. Whether this was insight, accustoming, war necessities, or the dieing out of old opponents, we do not know. In 1953, Staudinger was awarded the Nobel prize in chemistry after more than thirty years of work.

5 The Mysterious Crazy Glue

"Can you do addition," the White Queen asked,
"What's one and one and one and one and one
and one and one and one and one and one?"

L. Carroll, *Through the Looking Glass*

5.1 Macromolecules Step by Step

In the good old times, as homemakers were just housewives and did not enjoy the blessings of industrially manufactured foods, they would often prepare caramel. Sugar (more precise: saccharose or sucrose, commonly known as cane or beet sugar) was heated with a little water until it became brown and highly viscous. The brown color is due to decomposition products of the sugar, some of which produce the nice caramel taste whereas others cause cancer. But what increases the viscosity? Certainly a polymerization of the low molar mass sugar to higher molar mass compounds. But how does such a polymerization proceed?

Exactly the same question was asked in 1928 by the young American chemist Wallace Hume Carothers. He and some other promising young scientists were just hired by the E. I. DuPont de Nemours company in Wilmington, Delaware, in order to start a diversification program. The company let these researchers choose their own research topics, very revolutionary for those times and even today not exactly commonplace. Carothers knew the pro's and con's about the existence of macromolecules and he was convinced that long chains really existed. The synthesis of macromolecules compounds from low molar mass molecules was however rather mysterious. In the polymerizations studied by Staudinger, for example, the polymerization of formaldehyde to poly(formaldehyde),

$$n\ CH_2 = O\ \rightarrow\ \text{\small\sim\sim\sim} (CH_2 - O)_n \text{\small\sim\sim\sim} \tag{5-1}$$

or the polymerization of styrene to polystyrene,

$$n\ CH_2 = \underset{\underset{C_6H_5}{|}}{CH} \rightarrow \text{\small\sim\sim\sim} (CH_2 - \underset{\underset{C_6H_5}{|}}{CH})_n \text{\small\sim\sim\sim} \tag{5-2}$$

high molar masses were observed after the reaction of a few monomer molecules (low monomer conversions in chemical lingo). The molar mass did not change on further monomer conversion. Did this mean that many hundreds of monomer molecules were joined *simultaneously* to one polymer molecule? Did other laws as in regular chemistry apply where the simultaneous reaction of only three molecules to a new one is a very rare event? Questions, questions.

Table 7. Some early industrial polymers

Polymer	Dis-covery	Produc-tion	Typical applications
Thermoplastics from derivatives of natural polymers			
Cellulose nitrate	1846	1869	Knife handles, spectacle frames
Cellulose acetate	1865	1927	Photographic and packaging films, fibers
Fully synthetic thermoplastics			
Polyvinyl chloride	1838	1914	Shopping bags, window frames, artificial leather
Polyvinylidene chloride	1838	1939	Packaging films
Polystyrene	1839	1930	Small containers, foamstuffs, toys
Polymethyl methacrylate	1880	1928	Lamp casings, advertising signs
Polyethylene	1932	1939	Garbage bags, milk bottles
Thermosets from derivatives of natural polymers			
Casein/formaldehyde	1897	1904	Haberdashery
Fully synthetic thermosets			
Alkyd resins	1901	1926	Coatings
Phenol/formaldehyde	1906	1909	Electrical insulators
Natural rubber			
Natural rubber vulcanization	1839	1850	Tires
Synthetic rubbers			
Polyisoprene	1879	1955	Tires
Polybutadiene	1911	1929	Tires

Many semi-synthetic and fully synthetic polymers were already commercially used at that time (Table 7). The transformation of natural polymers into semi-synthetic polymers was fairly well understood in the late twenties since it employed known chemical reactions to already existing macromolecular compounds.

An example is the chemical transformation of cellulose into cellulose acetate. Cellulose is a polymer of glucose (raisin sugar) and thus a polysugar or polysaccharide. Like glucose itself, cellulose contains many hydroxyl groups. These groups can be transformed by acetic acid (vinegar) CH_3COOH to cellulose acetate; they are esterified to the acetic ester of cellulose

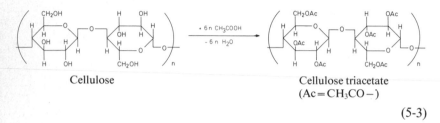

Cellulose   Cellulose triacetate
                                     $(Ac = CH_3CO-)$

$$(5-3)$$

This esterification reaction is the exact analog to the esterification of methanol (methyl alcohol, wood alcohol) with acetic acid to methyl acetate which was well known at the turn of the century

$$CH_3-O-H + HO-\underset{\underset{O}{\|}}{C}-CH_3 \rightarrow CH_3-O-\underset{\underset{O}{\|}}{C}-CH_3 + H_2O \quad (5-4)$$

methanol   acetic   methyl   water
                acid   acetate
                                (methyl ester
                                of acetic acid)

The skeleton of the cellulose chain is not changed during the esterification process, only the hydroxyl side groups react. How cellulose is made in nature was unknown at that time since mother nature has her own special and complex methods.

Carothers' idea was to apply the known chemical reactions of low molar mass chemistry not to side groups but to endgroups of small molecules. Each of these molecules must possess at least two reactive ("functional") groups if macromolecules were to result. Molecules with only one functional group react with each other only to give a "dimer" of two starting molecules as Eq. (5-4) shows for the formation of methyl acetate. If however a molecule with *two* hydroxyl groups (a glycol) is reacted with a molecule with two carboxylic groups (a dicarboxylic acid), then the resulting ester of two monomeric units will still have functional groups at each end

$$HO-CH_2-CH_2-OH + HOOC-CH_2-CH_2-COOH \rightarrow \quad (5\text{-}5)$$
Ethylene glycol $\quad\quad\quad$ Succinic acid

$$HO-CH_2-CH_2-O-OC-CH_2-CH_2-COOH + H_2O$$
Succinic acid monoethylene $\quad\quad\quad\quad\quad\quad\quad$ water
glycol ester (= ethylene
glycol monosuccinate)

The resulting simple ester can either react at the hydroxyl group with another succinic acid molecule

$$HO-CH_2-CH_2-O-OC-CH_2-CH_2-COOH$$
$$+HOOC-CH_2-CH_2-COOH \rightarrow$$
$$HOOC-CH_2-CH_2-CO-O-CH_2-CH_2-O-OC-CH_2-CH_2-COOH$$
$$+ H_2O \quad\quad\quad (5\text{-}6)$$

or at the carboxyl group with another ethylene glycol molecule

$$HO-CH_2-CH_2-O-OC-CH_2-CH_2-COOH +$$
$$HO-CH_2-CH_2-OH \rightarrow$$
$$HO-CH_2-CH_2-O-OC-CH_2-CH_2-CO-O-CH_2-CH_2-OH$$
$$+H_2O \quad\quad\quad (5\text{-}7)$$

The resulting trimers possess either two carboxylic endgroups as in Eq.(5-6) or two hydroxy endgroups as in Eq.(5-7). In each case, they are again bifunctional and may react with another molecule to give tetramers and further on to pentamers, hexamers ...

Each step generates molecules with an additional monomeric unit; each step leads to molecules with exactly known endgroups. After many of these steps, true macromolecules with exactly known chemical structures are formed, schematically

$$n\, HO-R-OH + n\, HOOC-R'-COOH \rightarrow \quad (5\text{-}8)$$
$$HO-R-O-(OC-R'-CO-O-R-O)_{n-1}OC-R'-COOH$$
$$+ (2\,n-1)\, H_2O$$

Carothers thus showed that macromolecules can be synthesized step by step by the known methods of organic chemistry if bifunctional monomers are used. Like a railway car, each molecule must at least possess two coupling links; otherwise no train of chain-like linked cars could be formed.

Especially successful was the reaction of hexamethylene diamine with adipic acid to poly(hexamethylene adipamide)

$$n \, H_2N-(CH_2)_6-NH_2 + n \, HOOC-(CH_2)_4-COOH \rightarrow \qquad (5\text{-}9)$$

Hexamethylene   Adipic acid
diamine

$$H[NH(CH_2)_6NH-OC(CH_2)_4CO]_nOH+(2n-1)H_2O$$
Poly(hexamethylene adipamide)

since long threads could be drawn from the melt of the resulting polymer. Carothers thus invented the first fully synthetic fiber. It was called "nylon" (more exactly: nylon 6.6 or polyamide 6.6) and became so famous that ladies' stockings made from polyamide 6.6 became simply known as "nylons".

5.2 The Art of High Conversions

Group after group is condensed in Reactions 5-8 and 5-9; these polymerizations are thus called condensation polymerizations or polycondensations. In principle, each endgroup A (OH or NH_2) can react with each other endgroup B (COOH) regardless of the number of interior groups. Successive addition of monomer molecules leads to dimers, trimers, tetramers, pentamers, hexamers, etc. The dimers can not only react with monomers but also with dimers themselves to give tetramers, hexamers, octamers, etc. The trimers may react not only with monomers and dimers but also with trimers, tetramers, etc. In short, groups react without regard of molecule sizes (see Eq. (5-10) on next page).

A nonamer may thus be formed from an octamer + a monomer, a septamer + a dimer, a hexamer + a trimer, or a pentamer + a tetramer. Since a group of a molecule reacts with equal probability with a group of a molecule of *any* other size, molecules with many different sizes exist at any point of the reaction (exclusive of the dimerization, of course). By no means do all monomers first completely react to dimers, then all dimers to tetramers, all tetramers to octamers, etc.

Polycondensations are thus statistical reactions and the numbers n in the polyester of Eq. (5-8) and the polyamide in Eq. (5-9) are thus *averages* over all molecule sizes. Polymer scientists call n (or $\bar{X}_n$ or $\langle X_n \rangle$) the number average degree of polymerization since it averages over the number of molecules which possess a certain degree of polymerization. Note that the n in Eqs. (5-8) and (5-9) relates to the average number of constitutional

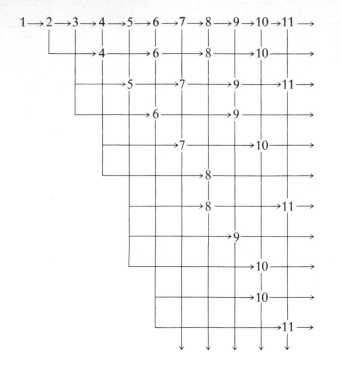

$$\text{(5-10)}$$

etc.

repeating units, not to the average number of monomeric units. There are two monomeric units per constitutional repeating unit in the examples of Eqs. (5-8) and (5-9).

The degree of polymerization is a theoretically important quantity which can be rarely measured directly. Experimentally, molar masses are determined. The two quantities are related to each other since the molar mass of the molecules is given by the sum of the molar masses M_E of the endgroups plus the product of the average number of repeating units $\overline{X}_n$ with their molar masses M_U:

$$\overline{M}_n = \overline{X}_n \cdot M_U + M_E \tag{5-11}$$

The polyamide 6.6 of Eq. (5-9) contains n repeating units $-NH-(CH_2)_6-NH-OC-(CH_2)_4-CO-$ and two endgroups (H and OH). The molar mass of such a molecule with a degree of polymerization (with respect to constitutional repeating units) of $n = \overline{X}_n = 200$ is thus $\overline{M}_n = (200 \cdot 226 + 17 + 1)$ g/mol $= 45\,218$ g/mol (see also Table A-4 in the Appendix). The molar mass of the endgroups (18 g/mol) can thus often be neglected against the molar mass of the constitutional repeating units (45 200 g/mol in our example).

Since the degree of polymerization is an average, the molar mass must be an average, too. Suppose we have a mixture of just two types of molecules, P and Q. The P molecules are assumed to have a relative molar mass of 20 000 and the Q molecules one of 100 000. There are N_P of the P molecules and N_Q of the Q molecules. The number average molar mass of this mixture is the sum of the products of number of molecules by mass of molecules, divided by the sum of the numbers. Assuming as an example $N_P = 5$ and $N_Q = 3$

$$\overline{M}_n = \frac{N_P \cdot M_P + N_Q \cdot M_Q}{N_P + N_Q} = \frac{5 \cdot 20000 + 3 \cdot 100000}{5 + 3} = 50000 \qquad (5\text{-}12\,a)$$

There are many other average molar masses besides the number average. Most important is the so-called weight average which averages over the masses of molecules instead of their numbers. The mass m'_i of *one* molecule of size i is given by its molar mass divided by the Avogadro number N_A (see Chap. 3.5), i.e., $m'_i = M_i / N_A$. The mass m_i of *all* molecules of this size is thus the mass of a single molecule multiplied by the number N_i of these molecules. The mass average $\overline{M}_w$ of molar masses (commonly called the weight average) for our special case is thus

$$\overline{M}_w = \frac{m_P \cdot M_P + m_Q \cdot M_Q}{m_Q + m_P} = \frac{N_P \cdot M_P^2 + N_Q \cdot M_Q^2}{N_P \cdot M_P + N_Q \cdot M_Q} = \qquad (5\text{-}12\,b)$$

$$= \frac{5 \cdot (20000)^2 + 3 \cdot (100000)^2}{5 \cdot 20000 + 3 \cdot 100000} = 80000$$

All synthetic macromolecular compounds and many natural ones consist of mixtures of macromolecules with different molecule sizes whether they have the same overall chemical composition or not. These macromolecular compounds are not molecular homogeneous, unlike low molar mass compounds.

The knowledge of the type and the width of the distribution of the molecular sizes is very important for many applications. Molecules with low molar masses may be toxic whereas true polymers are almost never toxic since they cannot be resorbed by the body. Small molecules must therefore be removed from polymers such as plastic soft drink bottles which may come into contact with food. Small molecules, on the other hand, may be excreted by the body, for example through the kidney whereas bigger molecules may reside forever in the body if the body's enzymes cannot break them down into smaller molecules. An example of such an effect is poly(vinylpyrrolidone) which was used extensively as blood plasma expander in World War II under the name PVP when not enough human blood was available for transfusions. This polymer compound undoubtly saved many lives. However, it had a relatively broad molar mass distribution, and the higher molar masses could not leave the body and so remained deposited with potentially harmful side effects. PVP is no longer admitted as blood plasma expander in peace time since the risk from the side effects is judged to be too great. In war time, that risk is of course small compared to the risk to lose a life and PVP is cherished because it remains stable over long times under adverse conditions, whereas blood itself has a shelf life of only a few weeks.

The promiscous reactions of the endgroups in bifunctional polycondensations lead to an interesting consequence for the change of the degree of polymerization with the conversion of endgroups. The degree of polymerization is a very important quantity since it controls the mechanical properties of a polymer. Take the example of lactic acid (milk acid) $HO - CH(CH_3) - COOH$, a syrupy liquid. The polymers of lactic acid are called polylactides; they have the chemical formula $H(O - CH(CH_3) - CO)_n OH$. Polylactides with small n are waxy masses but polylactides with large n can be drawn into threads which are used for surgical sutures. The threads slowly decompose in the body to lactic acid by a reversal of the polycondensation reaction. Lactic acid is not harmful since it is often generated by the body itself, for example, by the incomplete "combustion" of sugars that causes a muscular ache.

The slow degradation of sutures from polylactides and similar compounds is beneficial to patients since it eliminates the painful pulling of threads after surgery. It is not necessarily so beneficial for the medical profession since there will be one less visit by the patient and one less visitation fee. The "good" doctors therefore often prefer old-fashioned silk which does not decompose in the body. This is, of course, only good for the patient since it allows a better control of the healing process as patients are told. And aren't we patient . . . and trustful . . . and thankful . . .

But back to polycondensations. At the beginning of the reaction, we have N_M^o lactic acid molecules, each with one OH group. The number of initially present lactic acid molecules thus equals the number of initially present OH groups ($N_M^o = N_{OH}^o$). OH groups react only with COOH groups, not with other OH groups under the chosen conditions. After some time, some OH groups reacted and their number decreased to N_{OH} from N_{OH}^o. The number of reacted OH groups is thus $N_{OH}^o - N_{OH}$. The fractional extent of OH group reaction is

$$p_{OH} = \frac{N^o_{OH} - N_{OH}}{N^o_{OH}} = 1 - \frac{N_{OH}}{N^o_{OH}} \qquad (5\text{-}13)$$

Water is generated during the polycondensation and it is continuously removed. At a certain extent of group reaction, a number of monomer molecules is united in polymer molecules. The total number of reactant molecules (polymer and remaining monomer) is now reduced to N_M from N^o_M. On the other hand, all initially present monomeric units, $N^o_U = N^o_M$, are still in the system. The number average degree of polymerization of the reactant is thus given by the ratio of the number of all monomeric units to the number of all reactant molecules

$$\overline{X}_n = \frac{\text{number of units}}{\text{number of molecules}} = \frac{N^o_U}{N_M} \qquad (5\text{-}14)$$

Each reactant molecule carries one OH group; the number of reactant molecules must thus equal the number of remaining OH groups ($N_M = N_{OH}$) whereas initially $N^o_M = N^o_U = N^o_{OH}$. The degree of polymerization can thus be written as $\overline{X}_n = N^o_{OH}/N_{OH}$. The same ratio appears in Eq. (5-13) which is therefore written as

$$p_{OH} = 1 - \frac{1}{\overline{X}_n} \quad \text{or} \quad \overline{X}_n = \frac{1}{1 - p_{OH}} \qquad (5\text{-}15)$$

Equation (5-15) shows how the number average degree of polymerization varies with the extent of group reaction in polycondensations (Fig. 4). The degree of polymerization is only $\overline{X}_n = 2$ if 50% of the groups have reacted ($p_{OH} = 0.5$). Chemists are typically very proud if they achieve a 90% reaction. Polymer chemists can not be satisfied with such a result: at $p_{OH} = 0.9$, the degree of polymerization is still only $\overline{X}_n = 10$. Good fibers require however a degree of polymerization of at least $\overline{X}_n = 200$. The polycondensation must therefore pushed to at least $p_{OH} = 0.995$ (99.5% conversion of groups) which is quite a requirement.

Care has to be taken in the interpretation of the degrees of polymerization. A number average degree of polymerization of 2 does not mean that all molecules are dimers. Note that the degree of polymerization refers to *all* reactant molecules, including monomers, and that the polymerization reaction proceeds statistically with respect to the reaction of endgroups. The statistics of such polycondensations lets one calculate how many molecules of a certain degree of polymerization are present at a given extent of group reaction. Figure 5 shows this schematically for the poly-

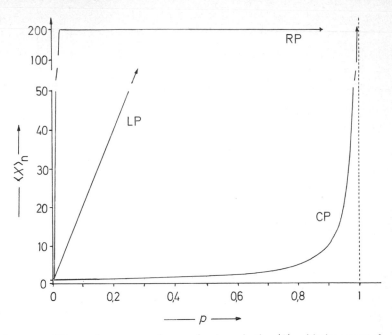

Fig. 4. Increase of the number average degree of polymerization $\langle X \rangle_n$ with the extent of reaction of functional groups p. RP = radical polymerization, LP = living polymerization, CP = condensation polymerization. The numbers are exact for bifunctional polycondensations of monomers with equal chemical reactivity but only schematic for radical and living polymerizations

condensation of 100 monomer molecules. At an extent of reaction of 50% ($p_{OH} = 0.5$; $\overline{X}_n = 2$) 25% of the monomer *molecules* did not react (but 50% of the *groups* did!). Ca. 13% of the molecules are dimers with consequently ca. 26% of all monomeric units (the exact figure is 25%). There are 6 trimer molecules, 3 tetramers, 1 pentamer etc. The fraction of polymer molecules with higher degrees of polymerization decreases rapidly; they are schematically represented in Fig. 5 (left) by one hexamer and one octamer.

The fraction of polymer molecules with higher degrees of polymerization increases with increasing average molecule size at the expense of the fraction of small molecules. At an extent of reaction of 90%, only 1% of the monomer molecules have not yet reacted. The distribution of polymer sizes is rather broad and can be schematically represented by one molecule each of trimers, pentamers, heptamers, nonamers, 11-mers, 14-mers, 19-mers, and 29-mers in our example (Fig. 5, right). In real life, molecules with all possible degrees of polymerization occur, i.e., also those with $X = 2, 4, 6, 8, 10, 12, 13$ etc.

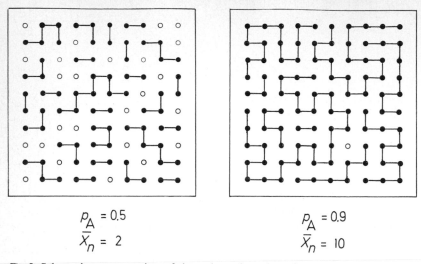

$$p_A = 0.5 \qquad\qquad p_A = 0.9$$
$$\bar{X}_n = 2 \qquad\qquad \bar{X}_n = 10$$

Fig. 5. Schematic representation of the polycondensation of A-B monomers. At 50% reaction of the A-groups ($p_A = 0.5$), 25% of the monomer molecules ($\bigcirc$) have not reacted to polymers ($\bullet\!\!-\!\!\bullet$). At an extent of reaction of 90% ($p_A = 0.9$), only 1% of the monomer molecules remain

High extents of reaction reduce the number of monomer molecules considerably. At 99.5% group reaction, only 0.003% (30 parts per million = 30 ppm) remain, at 99.9% only 1 ppm. These small amounts are generally below the monomer contents which are considered harmful to humans.

Purists will claim, however, that all these polymers are not "pure", since they still contain minute amounts of impurities such as monomers. And impurities are by definition harmful, aren't they? But which chemical substance, natural or synthetic, is absolutely pure? Every chemical always contains small amounts of "foreign" molecules. These amounts are often so small that they cannot be detected by present means or that they do not interfere with the intended use.

5.3 Macromolecules in One Stroke

Back to the old master Carothers. This gifted chemist not only invented nylon but also the first commercially successful synthetic rubber, polychloroprene. The chemical structure of polychloroprene is very similar to that of natural rubber. It carries a chlorine substituent instead of a methyl group whereas the so-called methyl rubber possesses two methyl groups

$$-(CH_2-C=CH-CH_2)_n- \qquad -(CH_2-C=CH-CH_2)_n- \qquad -(CH_2-C=C-CH_2)_n-$$
$$\qquad\quad | \qquad\qquad\qquad\qquad\quad | \qquad\qquad\qquad\qquad\quad |\quad | $$
$$\qquad\quad Cl \qquad\qquad\qquad\qquad\quad CH_3 \qquad\qquad\qquad\qquad CH_3\ CH_3$$

Polychloroprene   Polyisoprene   Polydimethylbutadiene
(Neoprene)   (Natural rubber)   (Methyl rubber)

Methyl rubber was produced in Germany during World War I, far before Carother's invention, since natural rubber was scarce because of the Allied blockade of the central powers. It was very cumbersome to manufacture. Big glass bottles were filled with the monomer and then exposed to the sun for three months. After this time, liquid monomer had been converted into solid methyl rubber. The polymerization rate was thus certainly not breath-taking. The rubber properties were also neither very reproducible nor comparable with those of natural rubber. Since the price for natural rubber reached extremely low levels after the war, methyl rubber was no longer produced.

The bad reproducibility of this polymerization and of other polymerization reactions continued to plague the industry. Sometimes the polymerization started immediately, sometimes hours passed before anything happened, sometimes the whole reaction exploded. Most production runs resulted in soluble polymers but under apparently similar conditions other runs gave intractable material which had to be removed from the polymerization reactor by "mining" that is, with hammer and chisel. Many years passed before chemists learned to control the polymerization of monomers with carbon/carbon double bonds like chloroprene $CH_2=CCl-CH=CH_2$ or styrene $CH_2=CHC_6H_5$. It took even longer to understand these reactions.

We know today that the polymerizations of ethylene derivatives

$$CH_2=CHR \xrightarrow{+R_I^*} R_I-CH_2-\overset{*}{C}HR \xrightarrow{+CH_2=CHR} R_I-(CH_2-CHR)_2^*$$
etc.
$$(5\text{-}16)$$

with e.g.   R = H for ethylene to polyethylene,
                CH_3 for propylene to polypropylene,
                Cl for vinyl chloride to polyvinyl chloride,
                C_6H_5 for styrene to polystyrene,
                $CCl=CH_2$ for chloroprene to polychloroprene, etc.
is initiated by certain active species R_I^* which can be radicals (R_I), cations ($R_I^\oplus$) or anions ($R_I^\ominus$). The active species add to a carbon/carbon double bond and generate a new active species which in turn adds more monomer molecules. In such addition polymerizations, only monomer is added to the growing chain whereas in condensation polymerizations all reactant molecules may add on.

The initiating species becomes covalently bound to one end of the resulting polymer molecule. The fraction of these endgroups is very small at high degrees of polymerization which lets us understand why earlier researchers did not find any endgroups at all.

But what is at the other end of the polymer chain? Well, this depends on the nature of the initiating species. Nomen est omen: Radicals are very radical and attack whatever they can find, especially those which do not resist (be they molecules or humans). The radicals themselves are destroyed during these reactions since a revolution eats its own children. Radicals also love fellow radicals and thus combine but here ends the analogy to humans since two combined radicals are quite dead and no longer aggressive

$$2 R_I(CH_2-CHR)_{n-1}CH_2-\dot{C}HR \rightarrow R_I(CH_2-CHR)_{2n}-R_I \qquad (5\text{-}17)$$

Radicals are generated, on the other hand, by the thermal cleavage of the electron pair of a single bond

$$R_I:R_I \rightarrow 2 R_I \qquad (5\text{-}18)$$

In free radical polymerizations, deliberately added initiator molecules $R_I - R_I$ are continuously split by heat to give initiator radicals which subsequently attack and add on to monomer molecules. The more initiator and polymer radicals are generated, the higher is the probability that these radicals combine with each other. Finally, an uneasy truce exists between the generation of new radicals and their elimination by pairwise combination: The overall radical concentration stays constant in this "steady state".

A growing polymer radical may "live" for a while and add monomer molecules before it is killed by combining with another radical. The number of added monomer molecules, i.e., the degree of polymerization, is obviously determined by the ratio of the rate of monomer addition to the rate of radical combination. In the steady state, both rates equal each other and the degree of polymerization becomes independent of the monomer conversion (Fig. 5).

Anions, on the other hand, possess a free electron pair (see, for example, the hydroxyl anion in Table 2). Anions cannot combine with other anions since this would violate the octet rule. Careful control of the polymerization conditions, especially the exclusion of all impurities, will thus keep the growing polymer anions "living". The discoverer of such "living polymerizations", Michael Szwarc, used the term "living" to indicate that the chains can "grow" if "nurtured" by monomer and are not "killed" by aggressive impurities or fellow polymers; the use of this term does not imply reproduction and metabolism.

In living polymerizations, the number of growing chains ideally is given by the number of initiator molecules. Each growing chain remains as such and each added monomer molecule increases the degree of polymerization by one. The number average degree of polymerization thus increases linearly with the extent of monomer reaction (Fig. 5).

Living polymerizations are industrially important for the synthesis of so-called block polymers which consist of long sections of unlike polymer units ("blocks") coupled together. One of the possible strategies to make triblock polymers is first to polymerize A-type monomers, then to add B-type monomers, and finally again A-type monomers

$$R_I^{\ominus} \xrightarrow{+\, n\, A} R_I A_n^{\ominus} \xrightarrow{+\, m\, B} R_I - A_n - B_m^{\ominus} \xrightarrow{+\, n\, A} R_I - A_n - B_m - A_n^{\ominus} \qquad (5\text{-}19)$$

The final anionic end is then destroyed. Several of these triblock polymers serve as so-called thermoplastic elastomers (see Chap. 12.4).

Addition and condensation polymerizations are rare in daily life. The formation of caramel, mentioned earlier, is an example but a very complex one. Much simpler is the chemistry of the title compound of this chapter. Chemically a so-called cyanoacrylate, $CH_2 = C(CN)COOR$, (plus certain additives), a drop of this liquid polymerizes like crazy in air in a few minutes. Adhesive joints from this material can bear weights of several thousand kilograms, hence the name "crazy glue".

It is not air (nitrogen, oxygen, carbon dioxide, noble gases) which causes the polymerization but the water vapor in the air. Water is always dissociated to a small amount into protons and hydroxy anions ($H_2O \rightarrow H^{\oplus} + OH^{\ominus}$). The hydroxy anions initiate the anionic polymerization of the cyanoacrylate. Since the surface of metals and other materials also contains initiating groups, resulting polymer chains are firmly anchored to the surfaces and the polymer produces a very strong adhesive joint.

Human tissue may also be glued together by cyanoacrylates. Morticians use it to glue lips and eye lids. Cyanoacrylates are furthermore used in surgery but only for such tissues where a certain necrosis (death of cells) can be tolerated, for example, kidney and liver operations but not heart surgery. Bleeding of wounds may be stopped by cyanoacrylates: The monomer polymerizes on the surface of the wound, forms a non-permeable polymer film, and thus stops further bleeding. The tissue around the wound is protected by a polyethylene film to which neither monomer nor polymer sticks. The polycyanoacrylate film is biologically decomposed after 2–3 months with the formation of formaldehyde, an antiseptic, which either oxidizes further to carbon dioxide and water or reacts with ammonia to give urea, all very harmless compounds at the levels involved.

Addition polymerizations also occur in nature. The beetle *Abax ater,* when in danger, shoots at its attacker a secretion which is mainly monomeric methyl methacrylate, $CH_2 = C(CH_3)COOCH_3$. The monomer polymerizes on the enemy and immobilizes him or her. We humans know the polymerized methyl methacrylate under such trade names as Plexiglas, Lucite, Perspex, and many others.

5.4 Giants and Their Functionalities

Why did early polymerizations often result in insoluble polymers? Well, the monomers were not pure enough but contained small amounts of difficult-to-remove impurities which "cross-linked" the polymer chains. A common impurity of styrene is, for example, the *p*-divinylbenzene

Styrene
(Vinylbenzene)

(*p*-Divinylbenzene)

p-Divinylbenzene contains two polymerizable carbon/carbon double bonds per molecule. Each double bond forms two single bonds on polymerization. *p*-Divinylbenzene is thus tetrafunctional whereas styrene is bifunctional. The polymerization of styrene itself leads to so-called linear chains (see Fig. 6 and later chapters). One double bond of *p*-divinylbenzene may be polymerized by one growing polymer chain and the other double bond by another chain. The monomeric unit of *p*-divinylbenzene thus links the two chains together (Fig. 6).

Simple macromolecules are converted into giant macromolecules by such a cross-linking process. On addition of sufficient *p*-divinylbenzene (yet relatively small amounts), all styrene molecules in a container are polymerized to a single molecule with an enormous molar mass. Suppose a 1-liter container is completely filled with macromolecules of the density of 1 g/mL and molar mass of 100000 g/mol = 100 kg/mol. This 1 kg of macromolecules corresponds to $(1\,kg)/(100\,kg \cdot mol^{-1}) = 0.01$ mol macromolecules. 1 mol always contains $6.02 \cdot 10^{23}$ molecules; we have thus $6.02 \cdot 10^{21}$ macromolecules of molar mass 10^5 g/mol. If all these macromolecules are interconnected, a single super-macromolecule of molar mass $(1000\,g) \cdot (6.02 \cdot 10^{23}\,mol^{-1}) = 6.02 \cdot 10^{26}$ g/mol results. Truly, a *real* macromolecule!

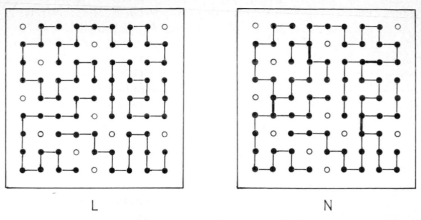

L   N

Fig. 6. Schematic representation of linear (L) and crosslinking (N) polymerizations.
○ Monomer molecules, ● monomer units in polymer molecules. —— Bonds between
monomer units in main chains, —— crosslinks between chains in networks

Cross-linked macromolecules are quite common in nature. Lignin, a main component of wood (see Chap. 9.3), is a highly cross-linked polymer whose chemical structure can be schematically represented by

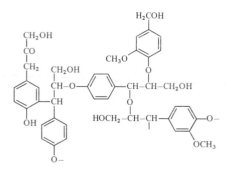

Would you think that coal is a cross-linked macromolecule? Well, not entirely, since bituminous coals contain both inorganic and organic components. The organic fraction consists of a soluble (uncross-linked) fraction of low molar mass aliphatic polymer chains and an insoluble, highly cross-linked fraction.

Cross-linking reactions are very often used to achieve special properties such as elasticity in cross-linked rubbers (Chap. 12.3) or thermal stability in thermosets (Chap. 11.1). The food industry makes also use of cross-links. Rice contains besides polysaccharides also proteins (see Table 18)

55

which possess some mercapto groups $-SH$ attached to long amino acid chains. These chains can be oxidized to sulfur/sulfur bridges

$$\text{\textasciitilde} S-H + (1/2)\,O_2 + H-S\,\text{\textasciitilde} \rightarrow \text{\textasciitilde} S-S\,\text{\textasciitilde} + H_2O \tag{5-20}$$

and form cross-links between the protein chains. The cross-linked proteins are less soluble and are less extracted by boiling water, which in turn increases the nutritional value of the rice and makes it more granular.

Soft wheat has less mercapto group containing amino acid residues in its protein molecules than hard wheat. Doughs from soft wheat are also less viscous than those from hard wheat and cannot be used for pasta where a kneading operation is required. Oxidation of the mercapto groups again increases the viscosity and generates a dough fit for pasta. Hard wheat, on the other hand, may have already too many crosslinks via sulfur-sulfur bridges so that the dough is to viscous. Some of these crosslinks can be removed by a reversal of the reaction of Eq. (5-20); the dough becomes less viscous so that a lot of energy is saved during the kneading or extruding.

Why are crosslinking reactions so easy to perform? The functionality of a molecule increases tremendously if molecules with functionalities higher than two are polymerized. The functionality of the resulting polymer molecules is always two if bifunctional monomers are used (Fig. 7), regardless of the polymer size. The functionality of a polymer molecule from monomer molecules with the functionality f_0 is however

$$f = 2(f_0 - 1) + (X - 2)(f_0 - 2) = 2 + X(f_0 - 2) \tag{5-21}$$

$$\qquad\text{terminal} \qquad\quad \text{mid-chain}$$

since two of the monomer functionalities are used by the $X - 2$ mid-chain units and one each by the two endgroups. The validity of this equation can be easily checked for, e. g., a hexamer ($X = 6$) of tetrafunctional ($f_0 = 4$) monomer molecules. Equation (5-21) applies to "linear" chains as in Fig. 7 B and to "branched" chains as in Fig. 7 C. It is no longer valid if groups within a molecule react under formation of intramolecular rings.

The functionality of polymer molecules from higher functionality monomers increases steeply with increasing degree of polymerization (Table 8). The pentamer ($X = 5$) of a tetrafunctional monomer has already a functionality of 12 and a 100-mer one of 202.

The resulting "networks" are not planar as apparently shown by Fig. 6. Since the carbon atom has a tetrahedral structure (Fig. 3), networks from carbon chain polymers must be three-dimensional in space. Bifunctional monomers, on the other hand, produce "linear" macromolecules which

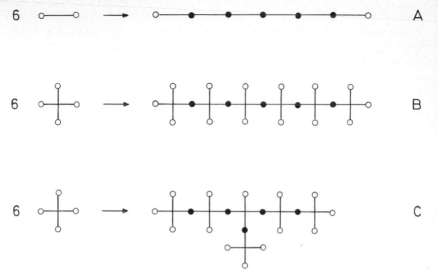

Fig. 7. Change of the functionality of polymer molecules upon polymerization of bifunctional (A) or tetrafunctional monomer molecules (B, C). Two functional groups (○) each form one chain bond (●)

may be "one-dimensionally" oriented. But what about "two-dimensional", platelet-like macromolecules? The functionality of their monomeric units must be at least three since otherwise no linking is possible in more than one dimension. If the bonds are not connected in a plane, then the growth of such molecules will be three-dimensional. How can one get a two-dimensional growth despite functionalities of three or more? Ask

Table 8. Influence of the functionality f_0 of the monomer molecules on the topology of the resulting macromolecules, excluding intramolecular ring formation. X = degree of polymerization

Functionality of the molecules of						Topology
Mono-mers X = 1	Di-mers 2	Tri-mers 3	Tetra-mers 4	Penta-mers 5	100-mers 100	
0	–	–	–	–	–	low molar mass
1	0	–	–	–	–	low molar mass
2	2	2	2	2	2	linear chains
3	4	5	6	7	102	networks
4	6	8	10	12	202	networks
5	8	11	14	17	302	networks

57

mother nature; she knows how to make mica and other platelet-like macromolecules from silicon ($f_o = 4$)[1], aluminium ($f_o = 3$) and some other chemical elements. We are unable to do that with carbon ($f_o = 4$) except for graphite, of course (see Chap. 14.3). Because of principle or because of insufficient knowledge? Who knows.

[1] For an introduction to the chemistry of silicon and its polyneric materials, see: E. Rochow: Silicon and Silicones, 1987 Springer Heidelberg New York.

6 Corn Syrup and Hi-Tech

6.1 Grass and Card-Board

American cows are sometimes fed paper and card-board reports the journalist and hobby farmer O. Schell in his book "Modern Meat". Their often pure corn diet does not contain enough fibers, and since fibers are good for digestion, paper and card-board fibers must help to overcome constipation. The TIME book reviewer read it in dismay and speculated that farmers certainly will soon feed plastics to the critters. How greedy: instead to feast on juicy natural grass, cows will have to munch on such unnatural materials like paper and card-board and the even less natural plastics. Nobody feeds plastics to cows, of course, and nobody will probably ever do it, but the specter of our plastics society is rising its ugly head again . . .

Or does it?

What is grass chemically and what is paper and card-board? Well, grass consists of cellulose, water, polyfructosans, chlorophyll and a number of salts and low molar mass compounds. Cellulose is the structural material; without cellulose, grass would be an amorphous mixture of its other components. Water makes the grass juicy, chlorophyll catalyzes the conversion of carbon dioxide into cellulose, polyfructosans, and other carbon compounds. Polyfructosans are the food reserve of grass and many other plants. They are formed from fructose (fruit sugar) by polymerization. The plants do not store fructose itself since fruit sugar is a low molar mass compound which diffuses rapidly through cell membranes. Polyfructosans are however macromolecular compounds which permeate cell walls only with difficulty or not at all; they are easy to store. In emergency situations, plants can rapidly degrade polyfructosans to fructose which is then easily transported through the plant. The degradation of high molar mass polyfructose to low molar mass fructose is a reversal of the polycondensation reaction: first bigger fragments are formed, then smaller ones, and finally fructose.

Cows love grasses with high polyfructose content: polyfructosans are easier to digest than cellulose and the resulting fructose is sweeter than the glucose from cellulose. Nature does not love the cows equally well, however. The polyfructosan content of grasses increases in the northern hemisphere up to ca. middle of May. The polyfructosan is then slowly converted into cellulose and the grass becomes stiffer and less juicy. Hay-making locks in the polyfructosan content and that is the reason why hay is made in May although the grass has not reached its full height. The grass also tries to replenish the lost parts and sprouts new growth.

Cows can digest cellulose although they like polyfructosans better. Paper and card-board, both mainly cellulose, are therefore not alien to them; they are natural foodstuffs for the cows. Why then the commotion? Because paper is man-made! How untrue, certain wasps can do that too.

"Unnatural" may be the printing ink in shredded newspapers but that is not true either. The main ingredient of black printing inks is carbon black (soot) which is naturally formed in any incomplete combustion of carbon-containing materials. Carbon black is very similar to charcoal and charcoal is used as medicinical coal against diarrhea. Do the cows know that?

Perhaps the cows but certainly not the many who are devoted to "organic" gardening. All naturally occuring macromolecular compounds are first decomposed in the soil to low molar mass compounds, then further to carbon dioxide, ammonia and water, typical "inorganic" materials. Plants never take up "organic" materials (i.e., complex carbon compounds) but only inorganic chemicals which was already known to Justus von Liebig 150 years ago. Naturally occurring guano and synthetic fertilizers are both inorganic materials. Only the solubility is different: synthetic fertilizers are often readily soluble in water and can be easily taken up by plants. Surplus fertilizer is however as easily washed into creeks, rivers, and lakes where it fertilizes algae and other undesirables. "Organic" fertilizers, whether it be leaves, coffee-grounds, or excrements, have first to be decomposed. The release of the inorganics contained therein is thus slow. The concentration of the inorganics available in "organics" is low which makes them less effective in the short run but may help in the long run (although manure around feeder lots of cattle is a major pollution problem!). It is the slow release and not the magic of the "organic" which works in organic gardening.

A slow release is also achieved by encapsulated synthetic fertilizers where the fertilizer grains are coated with a layer of a water-soluble polymer. The polymer slowly dissolves in the soil and the fertilizer is released in small dosages.

Macromolecular compounds are degraded to low molar mass compounds not only in dead materials. Every second, in every plant and animal, molecules are formed and destroyed. The average life-time of the macromolecular collagen of the connective tissue is many decades but human hemoglobin lasts only four months and the enzyme RNA-polymerase only a few hours. The decomposition of these macromolecular materials proceeds under the catalytic action of enzymes, naturally occuring macromolecular compounds.

Most synthetically produced polymers exhibit chemical structures which have no counterpart in nature. Evolution has thus not provided nature with enzymes which promote the degradation of many synthetic polymers. On the one hand, this is very desirable. Who would want clothes to degrade rapidly under the action of the enzymes of certain microorganisms as cotton did during jungle fighting in World War II. One the other hand, plastics litter is certainly not aesthetically pleasing. It is not harmful, though, unless one chokes on it but rusting metal cans and sharp-edged glass splinters are not more aesthetical either. But patience, patience, microorganisms are already learning how to decompose certain polymers. Evolution is still happening, even during a human life span.

6.2 Sweet Potatoes and Tough Meat

The helping hand of nature is not always desired. Who has not been grown angry over regular potatoes gone sweet? Potatoes are mainly composed of water and starch and starch is a mixture of the two macromolecular compounds amylose and amylopectin, both polymers of glucose. Amylose and amylopectin are water-soluble whereas cellulose, another glucose polymer, is not. The reason is a different primary bond between the glucose units (Fig. 8). This difference causes a different packing of the chains in the solid state which in turn causes different inter and intramolecular secondary bonding. Water may easily penetrate solid amylose and severe the secondary bonds, but not the more tightly packed cellulose chains. Amylose thus dissolves in water whereas cellulose is only slightly swollen and maintains its appearance.

Amylose and amylopectin have the same chemical structure except that amylopectin also possesses glucose units in side chains: amylopectin is a "branched" molecule. The ratio amylose/amylopectin differs in the various starches (Table 9). It is this ratio which makes the various starches useful for different applications.

Back to the sweetening of potatoes. At any time, two different chemical reactions occur in stored potatoes. Glucose molecules are polycondensed under liberation of water and, in a reverse reaction, amylose and amylopectin molecules are "hydrolyzed" by water to give glucose:

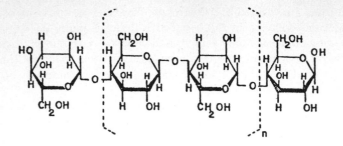

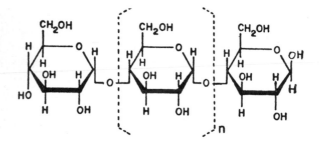

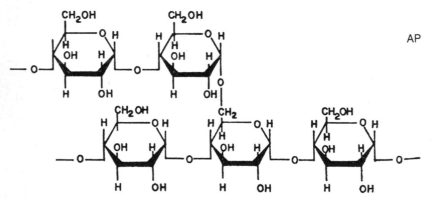

Fig. 8. Chemical structures of cellulose (C), amylose (A) and amylopectin (AP). Each chain consists of D-glucose units. The units are in so-called α-position in amylose and amylopectin but in β-position in cellulose. The repeating units of amylose and amylopectin are identical with glucose itself whereas the repeating units of cellulose consist of two glucose units. Amylopectin is branched, amylose and cellulose are not. Five of the six corners of the glucose "rings" are occupied by carbon atoms (not shown), the sixth by an oxygen atom

Table 9. Amylose content of various starches

Source	% Amylose	Source	% Amylose
Corn (hybrid amylomaize)	85	Pea (wrinkled)	66
Corn (regular)	26	Rice (regular)	19
Corn (hybrid waxy maize)	1	Rice (waxy)	0
Potato	22	Wheat	26
Tapioca	17	Oats	27

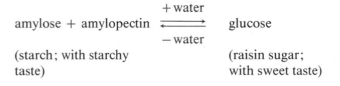

$$\text{amylose + amylopectin} \underset{-\,\text{water}}{\overset{+\,\text{water}}{\rightleftharpoons}} \text{glucose} \qquad (6\text{-}1)$$

(starch; with starchy taste)   (raisin sugar; with sweet taste)

The polymerization rate (reaction from right to left) increases strongly with temperature; it is very low at temperatures below 5 °C. The hydrolysis rate (reaction from left to right), on the other hand, depends very little on the temperature. At low temperatures, such as in a refrigerator, hydrolysis is faster than polymerization. More glucose is formed and the potatoes start to taste sweet. The sweetness can thus be removed if the glucose is polymerized. The polymerization rate must be higher than the hydrolysis rate for this purpose, which is true at higher temperatures. Sweet potatoes have thus to be stored for 2–3 weeks at temperatures above 22 °C and the sweetness disappears.

Starch is hydrolyzed industrially in great amounts. The United States produces each year 12 million tons of starch, mainly from corn, and Western Europe 10 million tons, mainly from potatoes, wheat and rye. The hydrolysis of corn delivers under certain conditions not only glucose but also the isomeric D-fructose. The resulting high-fructose syrup is very

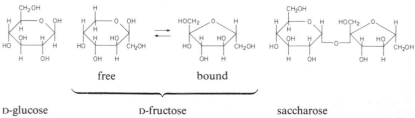

free   bound

D-glucose   D-fructose   saccharose
(raisin sugar)   (fruit sugar)   (cane sugar, beet sugar, "sugar")

63

sweet. Since it is also far less expensive than common sugar, it is used more and more to sweeten soft drinks, puddings, etc.

Not all polysaccharides can be hydrolyzed by the acid in the human stomach. Cellulose is an example; another one is the poly(galactomannan) of carob seeds. These seeds, by the way, have such uniform weight of 0.2 g that their weight became the unit with which precious stones such as diamonds are measured (1 carat; from Greek: keration, for the crescent-like shape of the carob seeds). Ground carob serves as nondigestible substitute for cacao.

Carob is an example of the "fiber" of food scientists. If these people speak of a "fiber-rich" diet, they do not mean that you should eat your last shirt or other fabrics from textile fibers. They also do not call "fiber" what remains after human digestion of plants. No, for them are "fibers" the leftovers after the successive treatment of defatted foodstuffs with a) boiling dilute acid and b) boiling dilute alkali. The "fiber" of food scientists may not have a fibrous form at all; it is a mixture of cellulose, pectins, hemicelluloses, and others.

Here is another example of depolymerization from genuine kitchen chemistry. A good restaurant owner should either have tender meat or sharp knives, at least in the old days. Today, even the most inexperienced cook can apply the discoveries of modern chemistry and convert the toughest meat into juicy steaks. How? Meat is mainly composed of proteins, fat and water. The proteins are macromolecules with alpha-amino acid residues

$$\sim\!NH-CHR-CO-NH-CHR'-CO-NH-CHR''-CO\!\sim \quad (6\text{-}2)$$

where R, R', R'', etc., are different groups, for example, H, CH_3, CH_2OH, CH_2CH_2COOH, etc. In aging animals, protein chains are intermolecularly crosslinked. Since the chains are now interconnected, they can no longer move freely: beef becomes tougher and humans less fit. In order to make steaks less tough, the network has to be split into smaller, more mobile, segments. Similarly to starch, this can be achieved by hydrolysis; for proteins by cleavage of the peptide bonds $-NH-CO-$.

$$\sim\!NH-CHR-CO-NH-CHR'-CO\!\sim + H_2O \rightarrow$$
$$\sim\!NH-CHR-COOH + H_2N-CHR'-CO\!\sim \quad (6\text{-}3)$$

Such a hydrolysis takes place on prolonged cooking of meat; as every homemaker knows, meat becomes less tough by this process. Unfortunately, the nice texture also disappears and we do want to charcoal-broil the steak and create a nice taste and many cancer-causing chemicals, don't we?

The hydrolysis may however also be accelerated by addition of certain enzymes. Examples are bromelin from fresh pineapple and papain from fresh papaya. Certain low-price steak houses know this very well and treat their steaks with the enzyme papain. The Haute Cuisine knows a better way: just add fresh pineapple and sell it as nouvelle cuisine. It has to be fresh fruit, however, since canned fruits are sterilized at high temperatures whereupon the hydrolyzing enzymes are destroyed.

6.3 Plastics Waste and Electronics

Degradation, like any chemical or physical process, is Janus faced. We want it for the production of corn syrup or the tenderizing of steaks but we do not want it for the sweetening of potatoes. Synthetic macromolecules are in this respect no different from biopolymers. Long lives are desired for plastic materials, textile fibers, rubber tires, surface coatings, or adhesives. Discarded tires, used packaging materials, and unfashionable clothings should however disappear as fast as possible.

Some polymers, like polyamides and polyesters, can be subjected to hydrolysis. Plastic soft drink bottles consist of a "polyester", chemically the same as in your polyester fabrics. After use, these bottles may be ground and hydrolyzed.

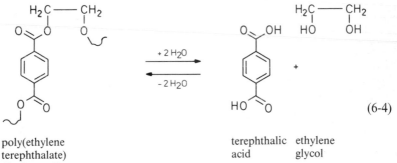

poly(ethylene terephthalate) ("polyester")

terephthalic acid   ethylene glycol

(6-4)

The resulting terephthalic acid and ethylene glycol may again serve as monomers for their polycondensation to polyesters (reaction from to right to left).

Modern computers would also not exist without depolymerization. Within forty years, room-sized machines were reduced to table-top instruments, thanks to transistors and integrated circuits. The manufacture of integrated circuits borrowed its technology from modern lithography which in turn depends on polymer science.

Photolithographic plates are mechanical printing devices. The desired images, for example, letters, are transferred from the plates to the paper by printing inks which are either on top of an elevated letter (positive or patrix) or in a hollow with the shape of that letter (negative or matrix). Truly, this is one of the few cases where as negative makes a positive impression which should console all those abhoring the chauvinism that a matrix (from Latin: matrix = womb (from mater = mother)) is always a "negative".

A printing plate is nothing but a collection of printing objects. Gutenberg was the first to arrange movable letters on a printing plate, at least in Western countries. This process is very slow and labor intensive and thus also expensive. In modern photolithography, a whole page is set up in type at once:

A plate is coated with a light sensitive polymer. The plate is then covered with a mask with the information (text, picture) and irradiated with ultraviolet light. Nothing happens to the covered parts. In the parts exposed to light a chemical reaction takes place between the light-sensitive groups at the polymer molecules. Since each polymer molecule contains several of such groups, many bonds are formed between the polymer chains and the macromolecules become crosslinked. The crosslinked molecules possess "infinite" size and are therefore insoluble in all solvents. The non-irradiated parts are not crosslinked and remain soluble. The non-crosslinked parts can thus be dissolved by a solvent whereas the irradiated parts remain as a relief and can be used as a printing device.

Integrated circuits are produced in a similar way. The "plates" are called "wafers" in integrated circuit manufacturing; they consist of highly purified silicon. The wafers are "doped" by certain chemical elements in the desired patterns; at these places, wafers become electrically conducting which enables them to act as circuits. The smaller the doped parts, the more can be arranged on a wafer, the smaller will be the computer at equal capacity.

The silicon wafers are first coated with silicon dioxide, $(SiO_2)_n$, a three-dimensional, regularly crosslinked polymer. Its microcrystalline variety is known as quartz, its macrocrystalline one as rock crystal. The "normal" symbol SiO_2 unfortunately does not reveal that silicon dioxide is really *poly* (silicon dioxide).

The silicon dioxide is then covered by a polymer which is sensitive towards ultraviolet light, X-rays, or electron beams. On irradiation, two

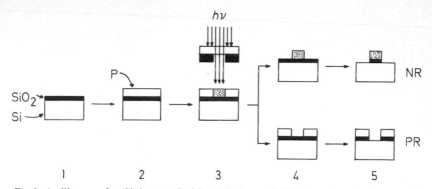

Fig. 9. A silicon wafer (Si) is coated with a thin layer of SiO_2 (step 1) and subsequently covered with a polymer P (step 2). The polymer is irradiated through a mask with the desired pattern (step 3). The polymer crosslinks under irradiation in so-called negative resists (NR); the remaining uncrosslinked polymer is dissolved (step 4, top). Irradiation decomposes the polymer to its monomers in positive resists (PR) (step 4, bottom). The SiO_2 is no longer protected by the polymer and is subsequently destroyed by etching (step 5). The exposed silicon may now be doped

possibilities exist. In one case, the polymer crosslinks at the irradiated places and the non-irradiated polymer is removed by a solvent (so-called negative resists) (Fig. 9). In positive resists, the irradiated polymer is not crosslinked but degraded to low molar mass compounds; the non-irradiated parts remain. In both positive and negative resists the silicon dioxide layer no longer covered by polymer is then removed by etching. The polymers have to resist this process, hence their name. The uncovered silicon parts are subsequently doped.

The doped bands should be very narrow with sharp edges. Ultraviolet light has wavelengths of 300–400 nm which allows to create bands of ca. 1000–2000 nm width. Much smaller structures can be created by X-rays (wavelengths of 0.1–1 nm) or electron beams (wavelengths, 0.05–0.005 nm). The dissolution of nonirradiated polymers in the negative resist process is also quite non-specific and does not generate the desired sharp edges. X-ray or electron beam irradiation and positive resists with their "dry" process are thus preferred for the manufacture of very large scale integrated circuits (VLSI).[1]

A suitable polymer is the polymer with alternating units from butene-1 and sulfur dioxide which is generated by copolymerization of the respective monomers

---

[1] More on chips in Rochow's book (see p. 184).

$$\begin{array}{ccc} CH_2{=}CH + S{=}O & \xrightleftharpoons[\text{depolymerization}]{\text{copolymerization}} & \text{\tiny vvv}(CH_2{-}CH{-}S{-}O)\text{\tiny vvv} \\ \quad | \qquad\quad \| & & \qquad\qquad | \quad\; \| \\ \quad CH_2 \quad\; O & & \qquad\qquad CH_2 \;\; O \\ \quad | & & \qquad\qquad | \\ \quad CH_3 & & \qquad\qquad CH_3 \qquad\qquad (6\text{-}5) \end{array}$$

butene-1   sulfur   poly(butene-co-
              dioxide   sulfur dioxide)

On irradiation, a chain bond may be broken in the polymer molecule. Contrary to many degradation processes discussed before, the two molecule fragments do not remain as such and are then cleaved again at random. Rather, one monomer after the other is split off the original scission position until the whole molecule is decomposed into its monomer components, butene-1 and sulfur dioxide. Each chain need be split only once; monomer molecules are generated in an unzipping reaction in this depolymerization, the true reversal of a polymerization (Fig. 10).

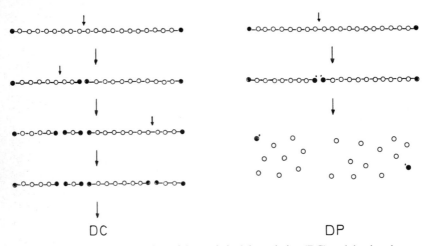

DC   DP

Fig. 10. Schematic representation of the statistical degradation (DC) and the depolymerization (DP) of a polymer. In both cases, a polymer chain is first broken at random which leads to two smaller chains with new endgroups (●) in the statistical degradation. The new chains are repeatedly degraded until all units are converted into monomers. The statistical degradation is the reversal of a polycondensation. The first step in depolymerization is the formation of two radicals ($-●\;^{.}$). The radicals are instable above a certain ceiling temperature which is characteristic for each polymer. Polymer chains are unzipped to monomer molecules above the ceiling temperature: The depolymerization is a reversal of the addition polymerization

68

7 Engine Oils and Vanilla Sauces

God does not roll dice.

A. Einstein

7.1 Leftists are Less Preferred

Have you ever asked yourself what the letters and numbers mean on engine oil cans? And why starch is such a good thickener for pudding sauces? Do you think that engine oils and pudding sauces have nothing in common?

To ask the question means to answer it. The effect of all pudding sauces and some engine oils is indeed caused by the same principle. Do you remember Chapter 4 where we discussed how Staudinger tried to prove the existence of macromolecular chains by viscosity measurements? And that he erroneously assigned them a rigid rod-like structure?

Look at a space-filling model of an ethane molecule (Fig. 11). Ethane with the chemical structure $H_3C - CH_3$ consists of two connected carbon atoms which are each substituted by three hydrogen atoms. Each carbon atom has four valences. The resulting four bonds do not reside in a plane but are regularly arranged in space (Fig. 3).

How do the three hydrogen atoms of the one carbon atom arrange relative to the three hydrogen atoms of the other carbon atom? Two extreme cases are possible (Fig. 11). In one case, each hydrogen atom of one carbon atom is situated directly above a hydrogen atom of the other carbon. They face each other and on the same side of an imaginary plane through the carbon atoms. Or, since scientists are fond of Latin, they are in "cis" position (if one is a polymer scientist). In the other case, the hydrogen atoms are "staggered"; they are in "trans" position (on opposite sides). The cis and trans positions are the two extreme "conformations" of the ethane molecule.

We know already that graphic representations often depict too small the atomic volume of the substituents. What happens if one lets the substituents grow and grow? At a certain size, substituents in cis-position would start to hinder each other whereas they would find easy accommodations in the trans position. Substituents in cis position thus try to reach the trans position which is energetically preferred over the cis position. All what is needed is a rotation of 60° about the carbon/carbon axis.

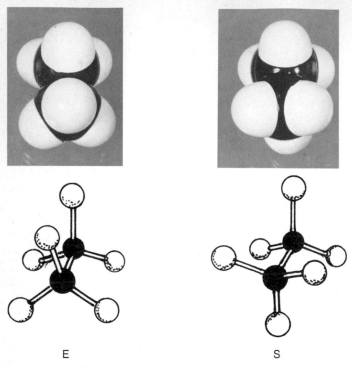

E   S

Fig. 11. Space filling models of ethane, H_3C-CH_3, with hydrogen atoms (white) and carbon atoms (black) in cis position (left) and trans position (right)

The ethane molecule is not so rigid that all hydrogen atoms are exclusively in trans position. The energy difference between trans and cis is just not big enough and a small kick is sufficient to convert a trans position into a cis position and from there, all by itself, back into trans. Where is that extra kick coming from? Each system takes up external heat and this energy is used by the molecule to reach the energetically less favorable cis position. It is like mountain climbing: to rest in a valley is comfortable but additional energy is needed to get to the top. And why would one want to climb a mountain? Because it is there, that's why ...

The situation is a little bit more complicated for a "chain" molecule with four carbon atoms, the butane molecule $CH_3-CH_2-CH_2-CH_3$. The two center carbon atoms of this molecule possess two hydrogen substituents and one methyl substituent each (Fig. 12). There are three possible cis positions and three trans positions. Since the methyl group is much bigger than a hydrogen atom, all cis positions are sterically hindred and therefore unlikely. We can thus concentrate on the trans positions.

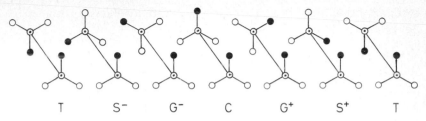

Fig. 12. Schematic representation of the six ideal conformations of butane, $CH_3-CH_2-CH_2-CH_3$, with methyl groups (●), carbon atoms (⊙), and hydrogen atoms (○). T=trans, S=skewed, G=gauche, C=cis

Two different kinds of trans positions exist (Fig. 12). In one of them, the two methyl groups are directly trans to each other whereas in the other two, the methyl group is trans to a hydrogen. It is easy to see that the former position is the energetically most favorable; only this position thus carries the true title "trans". The two other positions relate to each other like image and mirror image. In each, the methyl groups are off the absolute trans positions, they are "gauche" (French: gauche=left, akward, skewed). These two gauche forms are absolutely identical as one can see if one looks at the molecule from the opposite side. It would not make sense, therefore, to distinguish these two positions as "gauche" and "droit" (French: droit=right). Gauche-left and gauche-right would be a possibility. But since nobody has ever seen a leftist right, these positions are called gauche-minus and gauche-plus. Thanks heaven for foreign words: negative leftists would be truly disgusting.

7.2 Coils

Since conformations can be easily changed from trans into gauche and back, polyethylene-like chain molecules cannot be rigid rods. But what has this to do with engine oils and pudding sauces? Patience, dear reader, first we have to look at real chain molecules, say, polymethylene chains $H(CH_2)_nH$ with a degree of polymerization of $n=20000$. The molar mass of this chain is about $M \approx 20000 \cdot 14 \, g/mol = 280000 \, g/mol$. Polymers with this chemical structure are industrially produced as polyethylenes $\sim\sim\sim(CH_2-CH_2)_m\sim\sim\sim$. Their application comprises polishing waxes with molar masses of ca. 5000 g/mol, packaging films with molar masses of ca. 100000 g/mol, and ultra-high molar mass polyethylenes for artificial knee joints with molar masses of ca. 3000000 g/mol.

A polymethylene with a degree of polymerization of 20000 possesses 20000 chain bonds between the methylene groups. Each chain bond exhibits left and right one polymethylene substituent since such a chain can be written as

$$H-(CH_2)_p-CH_2-CH_2-(CH_2)_q-H$$

p and q may be arbitrary numbers between 1 and 19996 since $p+q+2=n$. Each polymethylene substituent can rotate around the valence angle between the chain bonds (Fig. 13). We just saw that only three conformations are preferred from the many possible: trans, gauche-minus, and gauche-plus. According to probability theory, such a polymethylene molecule can therefore adopt

$$3^{20000}=10^{9542}$$

possible macroconformations or shapes, that is a ten followed by 9541 zeros. The whole universe does not contain as many molecules as a single polymethylene molecule can adopt spatial positions!

A variety of spatial positions can be easily depicted by rolling a die. The six numbers on a die are assigned six directions in the paper plane (Fig. 14). We play the die thirty times which corresponds to a molecule with thirty chain bonds. For the first number, we place a bond in the corresponding direction, followed by the direction of the second number and so on. Whenever a number is followed by another one indicating the exact backward direction (for example, a six following a three), then that number is left out since a bond in 3-direction is excluded for a bond in 6-direction. The number sequence for the molecule in the upper left corner of Fig. 14 is thus 1-2-1-3-3-5-3-5-4-... The same procedure is then repeated for five other molecules with the same chain length.

The result is very educational. No molecule is a rod since this would mean an uninterrupted sequence of thirty times the same number (e. g., 4) or fifteen times the same binary sequence of two numbers (for example: 4-2). Not one molecule is "round" in the paper plane, that is, spherical in three-dimensional space. As was shown already by W. Kuhn in 1934, such

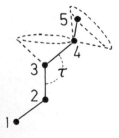

Fig. 13. Schematic representation of the rotation around chain bonds. Carbon atom 4 moves around a circle (broken line) at constant bond length 3–4 and constant valence angle τ; it resides preferentially in the trans, gauche-minus, and gauche-plus positions relative to carbon 1 (not shown). Carbon atom 5 moves around bond 4–5, etc.

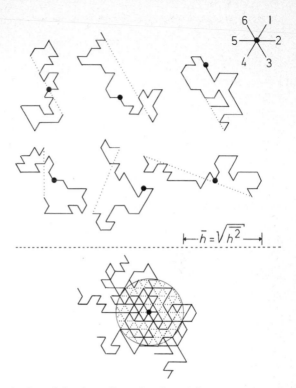

Fig. 14. Two-dimensional projection of the three-dimensional spatial arrangements of chains with thirty chain bonds each. End-to-end distances are indicated by dotted lines, the central chain atoms by a black dot. All six chains are projected on top of each other in the lower picture. The circle indicates the range of the radius of gyration. All chain segments outside the circle fit into the empty spaces (dotted lines) inside the circle

isolated macromolecular chains with more or less rotation around chain bonds are neither rods nor spheres but rather bean- or kidney-shaped.

One should not forget, however, that the shapes in Fig. 14 are snapshots of continuously changing shapes of a single molecule. Macroscopically, we also always look at very many molecules, never a single one. If thus all six molecules of Fig. 14 are projected onto each other on their centers of gravity, a picture for the spatial and temporal molecular shape results. This average molecular shape is much more rounded in two and more spherical in three dimensions. It resembles a loose coil and one is thus speaking of coil molecules. The coil shape can be clearly seen in electron micrographs if the chain diameter is large enough, for example, as in deoxyribonucleic acids (Fig. 15, see also Table 5).

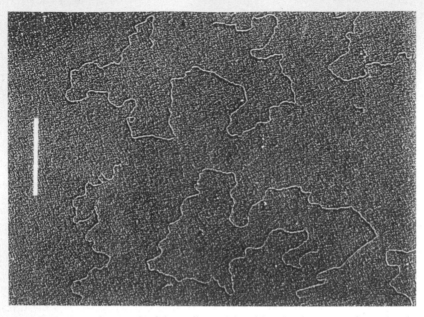

Fig. 15. Electron micrograph of deoxyribonucleic acid molecules on a surface. The picture corresponds to a two-dimensional projection of dissolved three-dimensional coils. The chains itself consist of so-called double helices (see Chapt. 8). After D. Lang, H. Bujard, B. Wolff, and D. Russell

The spatial dimensions of coil molecules are very small compared to their stretched-out lengths. Our polymethylene molecule with a molar mass of $280\,000$ g/mol contains $N_{bd} = 20\,000$ carbon/carbon bonds. Each bond is $L_{bd} = 0.154$ nm long. A totally stretched molecule is thus

$$L_{cont} = N_{bd}L_{bd} = 20\,000 \cdot 0.154 \text{ nm} = 3080 \text{ nm} = 0.00308 \text{ mm} \qquad (7\text{-}1)$$

long. This length is the true contour length since it corresponds to the contour of the molecule (Fig. 16).

The true contour length can never be achieved. The energy required to deform the valence angle τ to $180°$ from $112°$ would be so great that the carbon/carbon bonds break. A forked twig would break similarly if forced out of its natural angle between the forks. A polymethylene chain will thus never reach the full true contour length but at maximum only a length which is predetermined by the valence angle. Simple geometric considerations give for this length

$$L_{max} = N_{bd}L_{bd}\sin(\tau/2) \qquad (7\text{-}2)$$

74

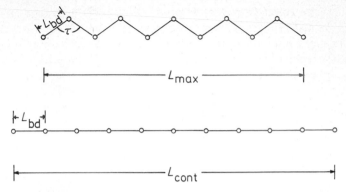

Fig. 16. Schematic representation of the maximum possible length (chain in all-trans conformation), L_{max}, and the true contour length, L_{cont}, of chains with the bond length L_{bd} and the valence angle τ. In carbon chains, a crystallographic length c may be calculated from a bond length of $L_{bd} = 0,154$ nm and a valence angle of $\tau = 112°$

and for our example $L_{max} = 2553$ nm (Table 10). Unfortunately, this length is often also called the contour length.

The maximum chain length may be achieved in certain polymer crystals but never in solution or in the melt since here the various carbon/carbon bonds reside in trans, gauche-plus and gauche-minus conformations: the molecule becomes coiled. The coiling can be measured by the radius of gyration or by the end-to-end distance of the chain ends. The end-to-end distance is of course only sensible for linear chains since branched chains possess more than two ends.

Table 10. Characteristic lengths of a poly(methylene) molecule $\sim\sim(CH_2)_n\sim\sim$ with a degree of polymerization of 20000 or a poly(ethylene) molecule$\sim\sim$ $(CH_2-CH_2)_n\sim\sim$ with a degree of polymerization of 10000

Type of length	Characteristic length in nm	
	theoretically calculated	experimentally found
True contour length	3080	–
Maximum chain length in all-trans conformation	2553	2553
Radius of gyration of an unperturbed coil	23	22
End-to-end distance of an unperturbed coil	57	54[a]
End-to-end distance of a coil with excluded volume (example)	–	63

[a] Calculated from radius of gyration.

The average radius of gyration for our polymethylene molecule is schematically depicted in the lower part of Fig. 14. It describes the radius of a spherical coil molecule with a homogeneous distribution of mass within the sphere. In reality, this cannot happen since some sites within this sphere are not occupied by chain segments whereas some segments reside outside the sphere.

End-to-end distances h and radii of gyration R for linear chains can be converted into each other as was especially shown by Paul J. Flory in statistical calculations

$$\overline{h_o^2} = 6\,\overline{R_o^2} \tag{7-3}$$

This equation was used to calculate an "average" radius of gyration, $R = \sqrt{R_o^2}$, from the average of the squares of the end-to-end distances of the six molecules of Fig. 14. Our polymethylene with degree of polymerization of 20 000 would accordingly have a radius of gyration of 23 nm and an end-to-end distance of 57 nm. The radius of gyration is much smaller than the maximum length; i. e., the molecules are strongly coiled.

The calculated radius of gyration of 23 nm refers to "unperturbed" coil molecules. Though molecule chains are very thin compared to their contour lengths, they are not infinitely thin; two chain segments can therefore not occupy the same site. We noticed that already as we rolled the dice where we excluded complementary numbers like 1 and 4, 2 and 5, or 3 and 6 (with respect to plane direction, not to faces of the die). Since two chain segments have to fight for the same site and each site is excluded to the other segment, coil molecules become perturbed relative to their ideal spatial arrangements; they are no longer "unperturbed".

The excluded volume is a direct consequence of the spatial requirements of chains of finite thickness, i. e., because of their steric repulsion. Some chain segments may however attract each other and will thus be able to overcome to some extent the effects of repulsion. The relative effects of repulsion and attraction depend on the solvent for the macromolecules since solvents also affect interactions. In certain solvents, effects of repulsion and attraction just cancel each other in such a way that the coil molecule behaves *as if* it were in the unperturbed state, that is, as if it would consist of infinitely thin chains. Complicated?

One should not picture such an unperturbed coil as a tightly packed globule, however. Coil molecules are rather loosely packed entities. Suppose an unperturbed coil with the average shape of a sphere and a diameter equalling the end-to-end distance. Our polymethylene molecule with a molar mass of $M = 280\,000$ g/mol and an unperturbed end-to-end distance of $h_o = 57$ nm $= 2\,r$ exhibits a volume of

$$V = (4\pi\, r^3)/3 = 96\,967\,\text{nm}^3 \approx 9.70 \cdot 10^{-17}\,\text{cm}^3$$

The mass of such a sphere is

$$m = M/N_A = (280\,000\,\text{g} \cdot \text{mol}^{-1})/(6.02 \cdot 10^{23}\,\text{mol}^{-1}) = 4.65 \cdot 10^{-19}\,\text{g}$$

and its average density therefore

$$\rho = m/V = (4.65 \cdot 10^{-19}\,\text{g})/(9.70 \cdot 10^{-17}\,\text{cm}^3) = 0.0048\,\text{g/cm}^3$$

If the average density of the solution is about $1\,\text{g/cm}^3$ (like water at room temperature), then only 0.48% of the coil volume is filled with polymer!

The chain units are not evenly distributed in dilute solutions, quite contrary to the behavior of low molar mass molecules (Fig. 17). Each chain unit is connected with two other ones; it cannot arbitrarily move to other sites. A (very) dilute polymer solution is thus made up of islands of polymer coils in a sea of solvent. Even within a coil molecule, chain units are not homogeneously distributed (see Fig. 14): the density is greatest in the interior.

Coil-forming chain molecules may swell in solution to multiples of their intrinsic volumes. Simultaneously, the viscosity increases dramatically. Why? Liquid water consists of water molecules with the small molar mass

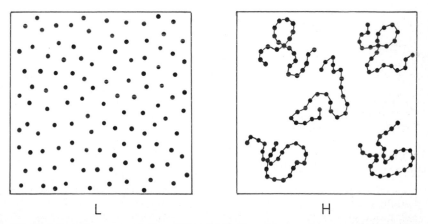

L   H

Fig. 17. Schematic representation of the distribution of low molar mass molecules (L) and high molar mass molecules (H) in solution. The total number of low molar mass molecules equals the number of monomeric units in high molar mass molecules in this example

of 18 g/mol. Very little energy is needed to shift the water molecules relative to each other: the viscosity is low. The whole big coil molecule (sometimes including the trapped solvent) must be moved in macromolecular solutions, however, and this needs plenty of energy: the solution is much more viscous, or, in common language, "thicker".

The secret of a good pudding sauce is just water-soluble macromolecules like amylose and amylopectin of starch or certain plant gums like guar or carrageenan. These macromolecules swell tremendously in water and increase the viscosity. The resulting "thickening" occurs not only in water but also in milk which is 87% water anyway.

7.3 Secrets of Engine Oils

The large space requirements of dissolved coil molecules are utilized in engine oils. Let us make a little detour to petroleum chemistry. Petroleum ("oil") is a complicated mixture of many hydrocarbons, i.e., chemical compounds consisting of carbon and hydrogen atoms. Crude petroleum often cannot be used directly, for example, because of its high viscosity. Since the high viscosities result from the too high molar masses of the hydrocarbon molecules, these molecules are first degraded into smaller pieces: they are "cracked" in the lingo of petroleum chemists. The first step of cracking is the random scission of molecules (see Fig. 10). Radicals

Table 11. Fractions from the cracking and distillation of petroleum in the United States and in Western Germany

Name of fraction	Number of carbon atoms in molecules	Boiling range in °C	Weight percent of fraction in	
			United States	Western Germany
Gas (refinery and liquid gas)	1– 4	<25	11.7	7.0
Naphtha (light gasoline and chemistry gasoline)	4– 7	20–100	55.0	19.0
Gasoline (transportation gasoline)	6–12	70–200		
Kerosine (heavy gasoline, aviation spirit)	9–16	175–275		
Gas oil (diesel oil, fuel oil)	15–25	200–400	24.9	38.7
Paraffin wax	18–35	230–300[a]		
Lubricating oil	25–40	300–365[a]	8.4	35.3
Bitumen	30–70	residue		
Petroleum coke	>70	residue		

[a] At reduced pressures between 0.07 and 0.10 bar.

are formed which combine with other radicals or attack other hydrocarbon molecules. The desired smaller molecules are only part of the newly formed molecules, the other part consists of undesired "crosslinked" ones (bitumen and petroleum coke).

The smallest of the originally present or newly formed hydrocarbon molecules are very volatile because of their small size; they form the refinery and liquid gas (Table 11). They are followed by naphtha, gasoline, kerosine, diesel and fuel oil, paraffin wax, lubricating oil, bitumen, and petroleum coke. The fractions are delivered in various proportions depending on the refinery design. If more gasoline is desired in the product mix, a more drastic cracking is required. This situation is typical for the United States with its high motorization and heating and air conditioning by natural gas and not by heating oil. Europe, on the other hand, needs relatively more heating oil and Europe's refineries thus use less cracking (Table 11).

Oil refining thus delivers a number of fractions. Some of the fractions may be used as engine oils, depending on their viscosities. The numbers found on oil cans (SAE 20, SAE 30, etc.) are a measure for the viscosities at $210°F$ ($=98.9°C$); SAE stands for "Society of Automotive Engineers" which introduced these norms.

The so-called single-range oils like SAE 20, SAE 30 etc. have one serious disadvantage: the strong temperature dependence of their viscosities. What may be "thin" at high engine temperatures, may be very "thick" in a cold motor and may thus cause cold-start problems.

Much more advantageous are multiple-range oils which cover more than one SAE viscosity class. These engine oils carry designations like SAE 10W-40. "W" stands for "winter". The number preceding the "W" indicates the temperature at which a certain viscosity prevails: the smaller the number the lower the temperature. Thus the larger the difference of the

Table 12. Properties of some engine oils

Property	SAE Grades						
	5 W-20	5 W-30	5 W-50	10 W-30	10 W-40	10 W-50	20 W-50
Dynamic viscosity in mPa·s at $-18°C$	1250			1250 –2500			5000 –10000
Addition of viscosity improver	no	yes	yes	no	yes	yes	no
Better running engine	+++	+++	++	+++	++	++	++
Easy cold start	+++	++	++	+	+	+	–
Energy savings	+++	++	+	+	+	–	–

two numbers, the larger the temperature range of good performance of the oil.

A good engine oil should have a low viscosity at 0° F for an easy cold start. At 210° F, it should not be too "thin", however. How to achieve that despite the strong decrease of the oil viscosity with temperature? Through addition of coil molecules!

These added coil molecules are known as VI (viscosity index) improvers. They permit the use of lighter base oils which have inherently a lower temperature dependence of viscosities. The strong effect of added polymers on oil viscosity can be seen if we rewrite the expression for the intrinsic viscosity (see Chapt. 4.4):

$$\eta = \eta_0 + \eta_0 [\eta] c + \ldots \tag{7-4}$$

where η_0 = viscosity of the solvent (here pure engine oil), c = concentration of the solute (here: added VI improver), and $[\eta]$ = intrinsic viscosity of the solute. If a polymer with an intrinsic viscosity of 240 cm^3/g is added to the oil in 1% concentration (0.01 g/cm^3), then the viscosity of the improved oil is 3.4 times higher than that of the pure oil! The reason for the strong increase in viscosity is the space requirement of the dissolved polymer molecules. Note that the intrinsic viscosity has the physical unit of a reciprocal concentration. An intrinsic viscosity of 240 cm^3/g means therefore that 1 g polymer occupies 240 cm^3 in solution whereas it needs far less space in the solid state (ca. 1 cm^3/g if the density of the polymer is ca. 1 g/cm^3).

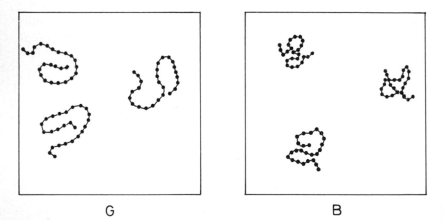

G   B

Fig. 18. Schematic representation of chains with the same degree of polymerization in a good solvent (G) and in a bad solvent (B). The chains have an excluded volume in the good solvent but are unperturbed in the bad one

The trick is to find coil molecules with low reduced viscosities at low temperatures and high reduced viscosities at higher ones. Or, in human terms: egocentrics which convert into outgoing individuals if activated by external stimulus. Self-loving egocentrics do have less contacts with the environment, they do not reach out that far (Fig. 18). But what makes coil molecules egocentric?

Let us compare methane CH_4 with water H_2O. Both molecules possess approximately the same molar mass (16 and 18 g/mol). Methane is a gas at room temperature whereas water is a liquid. Water molecules must therefore interact with each other more strongly than methane molecules: more energy is needed to separate and transfer them into the gaseous state. The reason is quite simple: water molecules contain two free electron pairs (Table 2) which interact with hydrogens from other water molecules. Water molecules are polar (Chap. 3) whereas methane molecules do not have free electron pairs and are apolar.

Polar interactions are much more temperature dependent than apolar ones. A viscosity-improving polymer for a good multiple-range engine oil certainly must have many apolar groups, otherwise it would not dissolve in the apolar engine oil (very polar molecules do not dissolve in apolar one: water and oil do not mix). The desired viscosity improvement with temperature is then achieved by the incorporation of a few polar groups into the polymer. At low temperatures, these groups like their equals in the same molecule; the coil is not expanded and the reduced viscosity is low. Increasing temperatures lower the readiness of the polar groups to interact just with equals. Contacts with the apolar molecules of the engine oil become more important. The coil molecule expands, the intrinsic viscosity increases, and the viscosity of the improved multiple-range oil becomes higher although the viscosity of the engine oil itself is reduced. To find the right balance between all these effects, that is the high art of polymer science . . .

8 Screwing Up Things

See plastic Nature working to this end,
The single atoms to each other tend,
Attract, attracted to, the next in place,
Form'd and impell'd its neighbour to embrace.

Alexander Pope (1688-1744), An Essay on Man,
Epistle III, Verse 9-12

8.1 Deterministic Coincidences

Staudinger was partially right after all: There are indeed rodlike macromolecules. Some even maintain their rodlike shape in solution. Other macromolecules resemble compact spheres. Yet, all the macromolecules investigated by Staudinger himself form neither rods nor spheres but only
ordinary coils.

Why? Coil molecules result if chain segments rotate more or less freely
around chain bonds (see Fig. 13). The successive directions of individual
chain bonds follow each other at random; they obey probability laws,
deterministic ones, that is. This sounds mysterious and even contradictary:
How can something be random if it follows deterministic rules?

Here we touch upon one of the great problems in the communication
between natural scientists and laypersons: the knowledge about the role of
chance in our world and the meaning of statistics. The throwing of a coin
is commonly seen as an example of a totally random process, just like the
rolling of a die. We cannot predict whether a coin will show head or tail.
Only the probability can be calculated and this probability is exactly 50%
for heads and 50% for tails. Whereby the use of the word "only" shows our
prejudice: Probabilistic statements are considered inferior to deterministic
ones by laymen.

Yet, the throwing of a coin or the rolling of a die is a deterministic process. If we know the exact spatial position of the coin before the throw, the
force with which the coin is flipped, the angular momentum we provide,
the frictional effects of air, the height of the fall, the gravitational force of
the earth and so on, we would be able to calculate the result of *each* throw.
Since we can never know all of these deterministic bases and since some of
them vary slightly with each throw, the flipping of a coin appears to us as

an apparently random process following probability laws. The process is random because of our ignorance; it is based on deterministic laws and thus reducible.

True random processes without deterministic bases are very rare. An example is the radioactive decay of instable atomic nuclei. Neither the laws of nature as presently known nor the initial state of a given atom allow us to deduce whether just this nucleus will decay next. The probability is not reducible to deterministic laws; it is an irreducible random process. The process is however not totally unordered or chaotic; it follows laws albeit probability laws, not deterministic ones. Human beings behave similarly since even our freespirited and anarchistic souls do not act totally chaotic but rather follow group dynamic laws.

The formation of macromolecular coils is quite similar. The directions of successive bonds follow each other statistically but only apparently at random since they are given by the chemical structure of the macromolecules and thus by the lengths of the chain bonds, the magnitude of the valence angles, the spatial requirements of the substituents, attractive and repulsive forces, and so on. Each chain segment has the freedom to arrange itself freely in space if structural dictates are not violated. How can one force such a freedom-loving molecule to give up its random ramblings and extend henceforth in only one direction? And how can one stabilize this direction against all external effects so that the overall shape of the molecule becomes rodlike?

Well, the same way one imposes law and order on chaotic members of our society: by self-discipline, by peer pressure, or by a hostile environment. Self-discipline is the most difficult of all since it rests on an inner disposition. It is also most difficult to achieve with macromolecules since it requires a chemical structure which is conducive to such "self-discipline". Since such structures result from an inner-self, they are often quite stable against external temptations.

Most simple is the forced orientation by peer pressure which ensures the party line in humans and chain alignments in polymer crystals. In humans and molecules alike, this kind of molecular alignment is disadvantageous if the lure of externals is to big: Humans defect to freedom-loving societies and crystallized molecules dissolve in certain liquids.

Hostile environments can also exert external forces on molecules but this will often lead to stocktaking of one itself and not to the assembling of like-spirited souls. Again, such order normally does not survive the removal of the external force.

8.2 Tactful Molecules

Let us look at some rod-like macromolecules. Such molecules may exhibit, for example, rigid chain units and stiff chain bonds without rotations around the chain axis. Self-disciplined molecules, so to say, with fairly stiff backbones. Poly (*p*-phenylene terephthalamide) molecules are such rigid macromolecules. Tradenamed Kevlar, they form fibers with excellent properties which are used for tire cord and body armor (see Chapt. 9).

In Kevlar, rigid phenylene groups $-C_6H_4-$ are connected in para-position via amide groups $-NH-CO-$ (Fig. 19). The chemical formula shown on the left side of Fig. 19 does not do justice to the true chemical structure since the chain bonds are not single bonds but partial double bonds as shown on the right side of that figure. These partial double bonds prevent segmental rotations around the chain bonds; the molecule chain remains stiff and rodlike. The rod-like character is inborn in the molecule, a typical case of self-discipline. Even the allurement of solvents does not change the rod-like character since the seducing solvent is forced to charge up the molecule (by "protonating" the amide bond with $H^\oplus$). The many positive charges along the chain repel each other and the molecule remains rod-like.

But look at our old friend, the polyethylene molecule (see Figs. 2 and 16). A zig-zag chain results if all carbon atoms occupy trans conformations (Fig. 20). Not much energy is needed, though, to transform a trans conformation into a gauche conformation. If the peer pressure by other chains no longer exists, then our molecule will take the opportunity and change itself from the extended all-trans conformation into a coil conformation with randomly arranged trans, gauche(+) and gauche(−) positions. Why? The many thousand chain atoms can adopt only one macroconformation as a zig-zag chain; they are highly ordered. In the coil conforma-

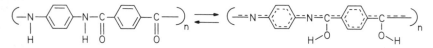

Fig. 19. Chemical structure of poly(*p*-phenylene terephthalamide). Dotted lines indicate "half" valence bonds which are distributed over the whole molecule. The resulting "one and one half" bonds stiffen the molecule and make it rod-like. The para-phenylene group is characterized by the symbol ⬡ (see also Chapt. 14.3). Kevlar melts under decomposition; it cannot be spun to fibers from the melt. It dissolves in concentrated sulfuric acid. The rod-like shape of the molecules is maintained in these solutions. Spinning from solutions creates a streaming gradient which lets the rods align parallel to each other, like tree trunks in a river. The orientation is maintained after removal of the solvent, giving strong, stiff fibers (see also Chapt. 11.2)

84

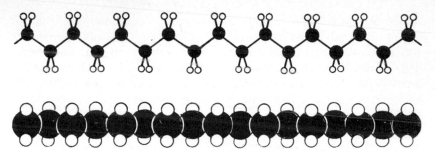

Fig. 20. Different models of a polyethylene chain. The size of the carbon (black) and hydrogen (white) atoms is not to scale

tion, many energetically-equal conformations are possible, however. A coil molecule is thus very disordered. You certainly know that all systems voluntarily attempt to reach a state with the greatest disorder (have you ever seen a teenager's room which cleaned up all by itself?). One thus has to force the polyethylene coils to adopt the highly ordered zig-zag form. How to do that? By peer pressure, that is, by other polyethylene chains. In polyethylene crystals, as we will see in Chapt. 11, polyethylene molecules are indeed in zig-zag form. The gentle peer pressure by other polyethylene chains is however removed in dilute solution; each chain freely unfolds to a coil.

Polyethylene is quite an "old" plastic. Researchers of the Imperial Chemical Industries company (ICI) in England tried in 1933 to compress gaseous ethylene under high pressure with a new machine from the Netherlands. They wanted to study the properties of ethylene under high pressure but the ethylene polymerized to polyethylene instead. Unfortunately, the polymerization was not reproducible; sometimes polyethylene was formed but sometimes not. A time-consuming search eventually located a very small crack in the apparatus. This crack allowed just the right amount of air to enter the reactor chamber; it was the oxygen of this air which started the polymerization. It took a lot of research to develop the industrial polymerization of ethylene: too little oxygen gave no polymerization. Too much oxygen and the whole plant exploded.

Many historians feel that neither generals nor soldiers won the Battle of Britain in World War II but polyethylene. This plastic is an excellent dielectric insulator for high frequency installations. Such a material was urgently needed for the newly developed radar which enabled the routes of German bombers to be determined, so as to alert and position British fighters. Without polyethylene no radar, without radar no early alert, without early alert no successfull defense.

Not all carbon chains prefer the all-trans conformation of zig-zag chains. Take polypropylene $\sim(CH_2-CH(CH_3))_n\sim$. Every second carbon atom of the chain carries here a methyl substituent CH_3. Polypropylene results from the polymerization of propene $CH_2=CHCH_3$ but for many years the polymerization turned out to be very difficult. Attempts to polymerize propylene free-radically with oxygen as with ethylene always gave oily low molar mass products and not solid plastics. The reason: strong branching reactions. Then, within two years, four different research groups found effective catalyst systems for the preparation of solid polypropylene. However, it then took almost thirty years before an American court declared the Phillips Petroleum company the winner in the subsequent struggle for priority rights for the composition of matter. And what a winner, since now all other polypropylene producers had to pay retroactive license fees. No small amount at an annual world production of about two million tons (two American-billion kilograms) and a sales price of ca. 80 US-cents per kilogram.

The polymerization of propylene became very popular after Karl Ziegler in Germany developed special catalysts for the so-called low pressure polymerization of ethylene. The Italian Guilio Natta then succeeded to polymerize propylene with these new catalysts to highly crystalline polypropylene. Both researchers received in 1963 the Nobel Prize in Chemistry for the discovery of this process.

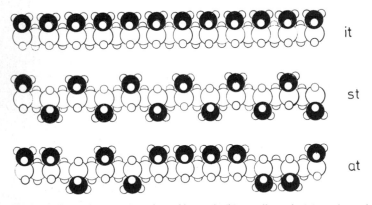

Fig. 21. Schematic representation of isotactic (it), syndiotactic (st), and atactic (at) polypropylene $\sim(CH_2-CH(CH_3))_n\sim$. Black balls = carbon atoms in methyl groups, big white balls = carbon atoms in the chain, small white balls = hydrogen atoms. Atoms and groups are not shown to scale. In reality, the big methyl groups force the chain to leave the pictured all-trans conformation and adopt alternating trans/gauche positions. The resulting helix is shown in Fig. 22

Ziegler's invention was caused by a systematic investigation of chemical reactions, lucky chance, clear realization of the cause of the chance observation, and again systematic investigations, in this sequence. History repeated itself since like the discovery of the free-radical polymerization of ethylene, a defect chemical reactor played a role. Ziegler did not intend to make polyethylene, he just wanted to prepare the scientifically and industrially interesting low molar mass aluminum alkyls $(R(CH_2CH_2)_i)_3Al$ (with $i = 1, 2, 3$, and so on) by a step-wise reaction of ethylene. One day, solid polyethylene was found in the reactor instead of liquid aluminum alkyls, but the polymerization was not reproducible. Some time passed before it was found that the reaction proceeded only in one of the many autoclavs. This particular reactor had a very small, hair-like crack and in this crack resided a little bit of a nickel compound as a left-over from previous experiments. The nickel compound together with the aluminum alkyls formed a catalytic system which in turn caused the polymerization of ethylene. Systematic investigation then led to catalyst systems which allowed the low pressure polymerizations of ethylene and propylene to crystallizable polymers.

The methyl groups can be arranged along the chains in two different ways (Fig. 21). They can be in the same tact, that is, isotactic (from the Greek: isos = equal, same; taktikos = ordered). The same methyl group position may also be reached only after two methyl groups; the chain is then in the syndiotactic configuration (Greek: syn = together; dio = two).

Polypropylene chains carry methyl groups on the first, third, fifth ... etc. chain carbon. The diameter of a methyl group is ca. 0.48 nm. The distances between the first and the third, the third and the fifth etc. chain carbon atoms are however only 0.254 nm (Fig. 16), much smaller than the minimum allowable distance between two methyl groups (their diameter). The methyl groups can thus not be accommodated if the chain is in all-trans conformation. The wise give in: The chain leaves the energetically unfavorable ... TTTT ... conformation and adopts the more favorable ... TGTG ... conformation which allows space for the methyl groups. In this ... TGTG ... conformation, successive bonds alternate between trans and gauche positions but all gauche positions must be either gauche (plus) or gauche (minus) (Fig. 22).

The methyl groups reside on a kind of spiral in these $(TG)_n$ conformations. The fourth, seventh, tenth ... methyl group is in the same relative spatial position as the first one. Three propylene units are thus necessary to complete a turn; one thus speaks of a 3_1 helix. The name helix originated from the spirally coiled shell of the snail Helix pomatia; but theirs is a conical helix, not a cylindrical helix as in isotactic polypropylene. The helix shown in Fig. 22 is by definition a left-winding one; it turns away counter-clockwise from the observer if viewed along the cylinder axis.

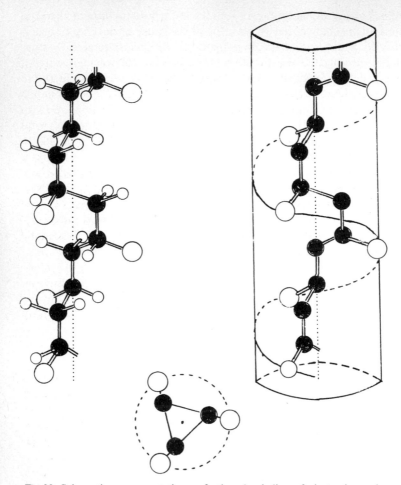

Fig. 22. Schematic representation of the 3_1 helix of isotactic polypropylene $\sim\sim(CH_2-CH(CH_3))_n\sim\sim$. White balls: hydrogen atoms (H), small black balls = carbon atoms (C), big white balls = methyl groups (CH$_3$). Left: spatial arrangement of atoms and groups. Right: same chain within a cylinder, not showing the hydrogen atoms. The positions of the methyl groups correspond to a left-winding helix since the screw turns away counter-clock-wise from the observer when viewed along the cylinder axis. The chain conformation is (TG$^-$)$_n$. Only the left-winding helix is shown. However, in isotactic polypropylene equal amounts of left and right winding helices are present

8.3 Helices and No End

You certainly heard before about helical macromolecules, probably through the nice book "The Double Helix" which describes the true life of a researcher (or at least that of its youthful hero, James Watson). This double helix is that of the deoxyribonucleic acids, the macromolecules of the genetic code. Some readers also know the so-called alpha-helix which appears in proteins. Both deoxyribonucleic acids and proteins are naturally occuring macromolecules and these are the models for the synthetic Johnny-come-lately's, aren't they?

Far from the truth! Helical structures were first discovered by none other than Hermann Staudinger and his coworker H.Lohmann at synthetic macromolecules (Table 13). Orienting X-ray measurements on crystallized polyethylene oxides $\sim(CH_2-CH_2-O)_n\sim$, replaced by more precise measurements of E.Sauter shortly thereafter, led Staudinger postulate a "meander" structure

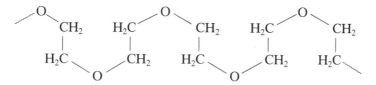

for the crystal structure of this polymer (after the old Greek river "Meander", now the Turkish "Menderes", which winds its way through the country side).

Such a meander structure is nothing but the two-dimensional projection of a three-dimensional helix into the paper plane! This feature was already recognized in 1933; Sauter could explain his results only by "strongly coiled meander chains" with "screw axes". Sauter also published space-filling models for polyoxymethylene, $\sim(CH_2-O)_n\sim$, which showed the typical helical structure.

These scientists did not use the magic word "helix", however. They rather talked in those years not only about meanders but also about spirals and screws. Was it the unfortunate choice of words that the importance of such structures was not recognized? "Helix" as a Latin-Greek word sounds much more scientific than "spiral" or "screw" and must therefore be more important and meaningful. Oh, the magic of words: sometimes one gets the distinct impression that a high-sounding terminology of a new phenomenon or idea is more important than the discovery itself . . .

The "coiling and uncoiling" of chromosomes was explained by Darlington in 1935 by the action of "invisible molecular spirals". Note that "coil-

Table 13. Theoretical proposals for and experimental determinations of helix structures

Macromolecular compound	Helix type	Theoretical proposal	Experimental determination
Polyethylene oxide	single	1932 H. Staudinger, H. Lohmann	1932 H. Staudinger, H. Lohmann
			1933 E. Sauter
Amylose	single	1937 C. S. Hanes	1943 R. Rundle, R. Baldwin, D. French
Polypropylene, isotactic	single	1942 C. W. Bunn	1953 G. Natta, P. Corradini
Poly(alpha-amino acids) and proteins	single	1943 M. L. Huggins	1951 L. Pauling, R. B. Corey
Deoxyribonucleic acids	double	1953 J. D. Watson, F. H. C. Crick	preceding experimental observations of various authors
Collagen	triple	1955 G. N. Ramachandran, G. Kartha; P. M. Cowan, S. MacGavin, A. C. T. North	ditto

ing" is not used here for the formation of random coils but for making and unmaking of spirals = helices. "Supercoils" and "coiled coils" are for biological scientists not specially coiled random coils but helices which are wound around each other. A "superhelix" is likewise in this jargon not a very perfect or a very long helix but the rope-like twisting of, for example, a triple helix. What Babylonian confusion! How was that with the White Queen in Chapter 1?

C. S. Hanes proposed in 1937 a spiral structure for the polysaccharide amylose in order to explain certain physical properties. The spiral structure was confirmed by X-ray measurements by R. Rundle and coworkers six years later, the first experimental proof of the helical nature of a natural polymer. Another crystallographer, C. W. Bunn, postulated shortly before that poly(α-olefins), $\sim\sim(CH_2 - CHR)_n\sim\sim$, must adopt the overall conformation of a helix if their configuration is regular (if they are isotactic in modern nomenclature). The polypropylene with $R = CH_3$, mentioned above, is the simplest representative of poly (α-olefins).

The „α-helix" of poly(α-amino acids) and proteins was the next helix to be discovered (Fig. 23). This helix possesses an additional structural principle as compared to the helices of polyethylene oxide, polyoxymethylene, and poly (α-olefins). The stability of the α-helix does not arise from the steric repulsion of the substituents R in the chains

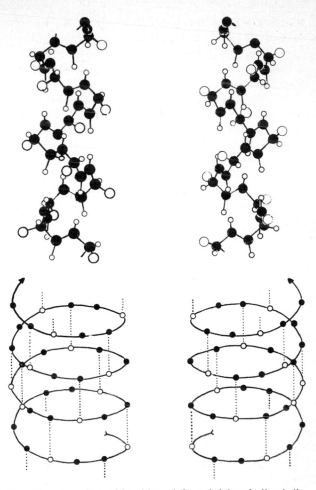

Fig. 23. Left-winding helix with D-α-amino acid residues (left) and right-winding helix with L-α-amino acid residues (right). Above: spatial position of atoms and groups. The chain atoms (in italics) $\wedge\wedge\wedge NH - CHR - CO\wedge\wedge\wedge$ are shown in black, the substituents R as big white balls, and the remaining substituents (H and O) as small white balls. Below: schematic representation of the helical conformation. Only chain atoms are shown (carbon in black, nitrogen in white). Hydrogen bonds $N - H \cdots O = C$ are represented as $\bullet \cdots\cdots \circ$

$\wedge\wedge\wedge(NH - CHR - CO)_n\wedge\wedge\wedge$ as in the chains of the poly(α-olefins) $\wedge\wedge\wedge(CH_2 - CHR)_n\wedge\wedge\wedge$ but from so-called hydrogen bonds $>N - H \cdots O = C<$ between two amide groups $\wedge\wedge\wedge NH - CO\wedge\wedge\wedge$ each (Fig. 23). It is this self-discipline of the intramolecular hydrogen bonds which keeps the α-helix together. The hydrogen bonds are so effective that they even survive the dissolution by certain solvents. An important difference to the helices of isotactic polypropylene which exist in the crystal state only since only here is the peer pressure big enough. What triumph of self-motivation and self-discipline!

The nice alpha-helices of poly(α-amino acids) and proteins are however only created if all α-amino acids have the same "configuration". The central carbon atom of an α-amino acid unit $\sim\sim NH - CHR - CO\sim\sim$ is surrounded by four different substituents, NH, H, R, and CO, where R may be the methyl group as in the alanyl residue or a variety of other groups. These four substituents arrange themselves in three-dimensional space on the corners of a hypothetical tetraeder (Fig. 3). Two possibilities exist if all four substituents are different. In one case, the substituents NH, CO and CH_3 follow each other relative to the substituent H in clock-wise direction, taking a right turn (Fig. 24). The constituent α-amino acid is thus called D-alanine (D from Latin dexter = right). In the other case, the sequence NH,

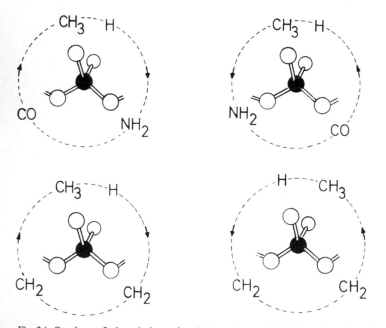

Fig. 24. Section of the chains of poly(α-amino acids) (above) and polypropylene (below). The central carbon atom is shown in black; the chemical structure of the substituents is given on the surrounding circle. Above left: unit of D-alanine, above right: L-alanine. Below: polypropylene (no asymmetry of immediate neighbors around the central carbon atom)

CO, CH_3 is arranged counter-clockwise, taking a left turn; this alpha-amino acid is called L-alanine (from Latin laevo = left). L-alanine and D-alanine relate to each other like image and mirror image or like a left hand and a right hand. These structures are thus also called "chiral" (from Greek: cheir = hand).

All amino acid units in a poly(L-α-amino acid) chain are relative to each other in the same position: the chain is isotactic. Contrary to isotactic polypropylene, each central carbon atom of isotactic poly(α-amino acids) is, however, immediately surrounded by four different substituents (see Fig. 24). The chain thus obtains a preferential direction which subsequently leads to the preference of one of the two possible helix directions over the other. Poly(L-α-amino acids) form only right-winding helices and poly(D-α-amino acids) only left-handed ones (Table 14). The helices of isotactic polypropylene, on the other hand, show no preferential direction; left and right-winding helices are always present in equal amounts. If the substituent R in poly(α-olefins), $\sim\sim(CH_2-CHR)_n\sim\sim$, is itself asymmetric, however, only then is the helical chain forced into a preferential direction.

Table 14. Helix types of various macromolecules

Macromolecules		Number of units per turn	Winding direction
Type	Example		
Poly(α-olefins) $\sim\sim(CH_2-CHR)_n\sim\sim$	isotactic polypropylene with R = CH_3	3	left and right in equal amounts
	isotactic poly-(3-methyl-1-butene) with R = $CH(CH_3)_2$	4	left and right in equal amounts
	isotactic poly-(5-methyl-1-heptene) with the asymmetric substituent R = $(CH_2)_2-CH(CH_3)(C_2H_5)$	3	left
Poly(α-amino acids) and proteins $\sim\sim(NH-CHR-CO)_n\sim\sim$	L units in α-helix	3.6	right[a]
	D units in α-helix	3.6	left
Poly(D-saccharides)	various	various	left[b]
Deoxyribonucleic acids	various in natural B-form	10	right

[a] Exception: poly(L-β-benzyl aspartate) with R = $CH_2-CO-O-CH_2-C_6H_5$.
[b] Several exceptions.

The L and D configurations of alanine look like image and mirror image and are thus normally considered energetically equivalent. The violation of parity in the subatomic range causes however a small energy difference in the two configurations which amounts to about 10^{-17} kJ/mol for alanine. This very small energy difference is of no consequence for our usual experiments and considerations. It is however significant for geological time ranges during which nature had time enough to weed out the thermodynamically less stable D-amino acids by kinetic means. All naturally occuring α-amino acids are thus of the L-type.

The double helices of the deoxyribonucleic acids (DNA) also have a preferential direction since each single chain has a preferential direction, too (Fig. 25). These double helices are especially stable. On one hand, two single chains are wound around each other; this alone creates an enhanced stability similar to that of a rope from two twisted single strands. On the other hand, the two strands are not only held together by hydrogen bonds between the bases but in addition by interactions between the base residues which are stacked one upon each other.

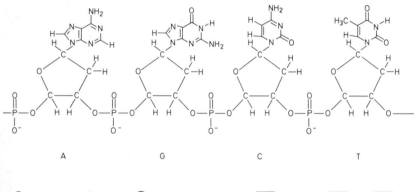

Fig. 25. Deoxyribonucleic acid. Above: chemical structure of a single strand of deoxyribophosphate chains with the attached bases adenine (A), guanine (G), cytosine (C) and thymine (T). Below left: space-filling model of a double helix with two single DNA strands. Below right: schematic representation of a double helix with base pairs

94

Helices and double helices are very important for the chemistry of life processes. Firstly, because of the division and multiplication of deoxyribonucleic acids which precedes the division and multiplication of living cells. Secondly, because of the strengthening and stiffening of the chains. Thirdly, and that is less known, because the helix formation is a very economic method to accommodate large quantities of macromolecules in very small spaces.

The bacterium *Escherichia coli* (from Greek: bacterion = small rod) has rod-like cells of 4 μm length and 1.4 μm diameter. The resulting volume of $6.2 \cdot 10^9$ nm^3 is home for about $4 \cdot 10^{10}$ molecules wherefrom 4 DNA molecules with an average molar mass of the DNA molecules of $2.5 \cdot 10^9$ g/mol and an average molar mass of monomer units of $M_U = 306$ g/mol (Table 15). A single DNA molecule in the random coil conformation would occupy a volume of at least $38 \cdot 10^9$ nm^3 (radius of gyration of 2080 nm). The space required by a random coil DNA molecule is thus much greater than the total space of the cell! A DNA molecule in the conformation of a double helix with a diameter of 2 nm (see Table 5) would however only occupy $2 \cdot 8.8 \cdot 10^6$ nm^3 = $1.76 \cdot 10^7$ nm^3, that is, only 0.3% of the available cell volume. Truly a fine, space-saving method. Nature stores food reserves the same way, for example, the partially helical amylose as the reserve polysaccharide of plants, the double chains of carrageenan as

Table 15. Approximate composition of a cell of the bacterium *Escherichia coli*

Molecule type	Number of molecule types	Average molar mass in g/mol	Number of molecules	Mass fraction in %
DNA	1	2 500 000 000	4	1.11
25 S r-RNA	1	1 000 000	$3.0 \cdot 10^4$	3.32
16 S r-RNA	1	500 000	$3.0 \cdot 10^4$	1.66
t-RNA	60	25 000	$4.0 \cdot 10^5$	1.11
m-RNA	1000	1 000 000	$1.0 \cdot 10^3$	0.11
Nucleotides	200	300	$1.2 \cdot 10^7$	0.40
Proteins	2500	40 000	$1.0 \cdot 10^6$	4.42
Amino acids	20	100	$3.0 \cdot 10^7$	0.33
Lipids	50	750	$2.5 \cdot 10^7$	2.07
Carbohydrates	150	200	$2.0 \cdot 10^8$	4.43
Other organic molecules	250	150	$1.7 \cdot 10^7$	0.28
Inorganic ions	20	40	$2.5 \cdot 10^8$	1.11
Water	1	18	$4.0 \cdot 10^{10}$	79.65
Total	4354		$4.05 \cdot 10^{10}$	100.00

reserve mucopolysaccharide of algae, and the poly(β-hydroxybutyrate) as the reserve polyester of bacteria.

Nature also uses helix formation in combination with coil-like segments to create almost spherical molecules. Some tricks are needed to do that. Examples of sphere-like molecules are the enzymes (from Greek: enzumos = sour dough), a catalytically acting class of proteins (from Greek: protos = the first, original). Proteins are naturally occuring copolymers of about twenty different types of α-amino acids. The amide bonds between their amino acid residues are called peptide bonds. Protein chains can also form helices although the substituents are not all alike as in poly(α-amino acids) but are different. These substituents should not prevent the helix formation, however. Since the α-helix is stabilized by hydrogen bonds, substituents should not contain groups which compete with hydrogen bond formation (Table 16). Such substituents would prevent helix structures and would rather generate coil-like segments.

Table 16. Naturally occuring α-amino acids $NH_2 - CHR - COOH$ and their ability to form helices or coils in poly(α-amino acids) (PAS) or proteins (PRO). Proteins may also contain in addition residues of the two imino acids

proline   and hydroxyproline

Name of amino acid	Substituent R	Name of amino acid	Substituent R
Helices in PAS and PRO		*Coils in PAS and PRO*	
Alanine	CH_3	Glycine	H
Phenylalanine	$CH_2 - C_6H_5$	Serine	$CH_2 - OH$
Leucine	$CH_2 - CH(CH_3)_2$	Cysteine	$CH_2 - SH$
Glutamic acid	$CH_2 - CH_2 - COOH$	Asparagine	$CH_2 - CO - NH_2$
Methionine	$CH_2 - CH_2 - S - CH_3$	Threonine	$CH(OH) - CH_3$
Lysine	$(CH_2)_4 - NH_2$		
Histidine		*Coils in PAS, helices in PRO*	
		Valine	$CH(CH_3)_2$
		Isoleucine	$CH(CH_3) - CH_2 - CH_3$
		Glutamine	$CH_2 - CH_2 - CO - NH_2$
Tryptophane			
		Unknown (varying with PRO)	
Helices in PAS, Coils in PRO		Ornithine	$CH_2 - CH_2 - CH_2 - NH_2$
Tyrosine	$CH_2 - C_6H_4 - COOH$	Hydroxylysine	$CH_2 - CH_2 - CHOH - CH_3$
Aspartic acid	$CH_2 - COOH$	Arginine	$CH_2 - CH_2 - N = C(NH_2)_2$

An enzyme molecule with such amino acid units contains stiff segments of rod-like (helical) portions which are held together by flexible segments of coil-like parts (Fig. 26). The stiff segments can be folded around the flexible ones. The resulting outer shape of the enzyme molecule is more-or-less spherical; segments are fairly tightly packed in the interior. The whole shape is then often stabilized by additional $-S-S-$ bridges between different segments.

Spherical molecules are very handy as transporting agents since spherical solute molecules create by far the lowest viscosity of solutions (Table 17), that is, the lowest resistance against flow. The transport of spherical molecules thus requires far less energy than the transport of rods or coils. All proteins with transport functions are therefore spherical molecules, for example, the hemoglobin for the transport of oxygen in the blood. Hemoglobin is a cluster of four myoglobin molecules (Fig. 26).

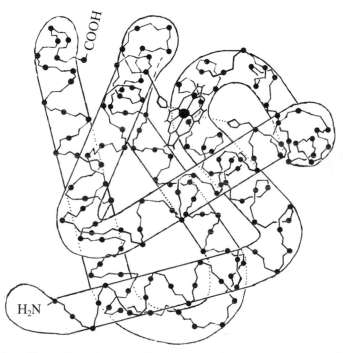

Fig. 26. Schematic representation of the protein myoglobin. Only the central carbon atoms (in italics) of the α-amino acid residues $-NH-CHR-CO-$ are shown; the other chain atoms occupy the corners of the connecting lines. Helical segments with the external shape of a rod are clearly seen to be connected by non-helical coil-like segments. The iron atom of the porphyrin rest is indicated by a big black point

Table 17. Comparison of relative and reduced viscosities of rods, coils and spheres. The reduced viscosity $\eta_{red}=(\eta-\eta_0)/(\eta_0 c)$ measures the spatial requirements of molecules, the relative viscosity $\eta_{red}=\eta/\eta_0$ gives the increase of solution viscosity over solvent viscosity. All data at a solution concentration of $c=0.01$ g/mL ($=$ "1%")

Molecule shape	Example	Molar mass in g/mol	η_{red} in cm^3/g	η_{rel}
Rod (as α-helix)	poly(γ-benzyl-L-glutamate)[a]	1 000 000	3000	31
Coil (unperturbed)	ditto[b]	1 000 000	120	2.2
Sphere (solvated)	Catalase[c]	250 000	3.9	1.039
	Hemoglobin[c]	68 000	3.6	1.036
	Myoglobin[c]	17 000	3.1	1.031
Sphere (unsolvated)	Dispersions and latices[d]	various	2.63	1.026

[a] In N,N-dimethylformamide, [b] in dichloroacetic acid, [c] in dilute aqueous salt solutions, [d] equal densities of spheres and solvent.

8.5 Foaming Spaghetti and Bursting Eggs

The helical segments of proteins are relatively easily destroyed as one can see almost daily in the kitchen. Haven't you seen spaghetti foam on cooking, or an egg burst and the escaping egg white curdle? In both cases, proteins are the culprits (Table 18).

Table 18. Composition of some foodstuffs

Foodstuff	Composition in %					
	Carbohydrates		Proteins	Lipids, fats	Water	Ash
	Fibers	Others				
Animal						
Eggs	—	1	12	11	75	1
Fish (flounder)	—	—	18	2	79	1
Fish (salmon)	—	—	18	6	75	1
Beef (lean)	—	—	20	11	69	1
Beef (hamburger)	—	—	16	28	55	1
Milk (Holstein)	—	5	3	3	88	1
Vegetable						
Soy bean	5	15	45	18	12	5
Oat meal	11	59	12	5	9	4
Wheat (winter)	3	69	14	2	10	2
Rice (whole grain)	9	65	8	2	11	5

Heating during cooking forces these proteins to leave their natural ordered conformation and to "denature" to a less-ordered one. In the natural conformation of these protein molecules, α-helical regions are held together internally by hydrogen bonds (Fig. 23). Added heat "melts" the hydrogen bonds and the formerly helical regions now adopt coil-like structures. This dissolution of a previous compact structure exposes "hydrophobic" groups which were buried in the interior of the protein molecule. The groups do not like the water, hence their name, and try to reach the more comfortable environment of their peers. The steadily added heat prevents them from returning to the beloved helical state and, out of desperation, they grab at random the surrounding peers. No longer

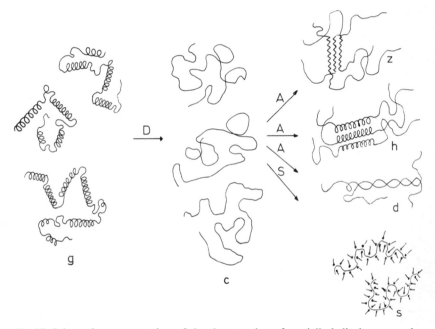

Fig. 27. Schematic representation of the denaturation of partially helical macromolecules, for example, spherical protein molecules (globulin g). The denaturation (D) dissolves the helical regions and random coils (c) are formed. Coil segments may associate (A) intermolecularly with other coil segments and generate a physical network PN. The association may occur via lateral alignments of zig-zag chains (z) or helices (h), or via formation of double helices (d), depending on the chemical structure of the macromolecules. No physical networks are formed in aqueous solutions if amphoteric compounds (chemical compounds with both hydrophilic and hydrophobic groups) are added to the solution (S), for example, vegetable oils to protein solutions. These additives solvatize the polymer chains (s) and prevent the intermolecular association

have these groups to rely on the attraction from groups of the same molecule, any peer from any other molecule will do. A physical network with coil-like segments is created which is insoluble in water and precipitates. The egg white curdles and the spaghetti foams.

How can one prevent the foaming? Since it obviously is caused by the association of hydrophobic groups, similar partners have to be added in great excess. Given the opportunity and being not very discriminating, the hydrophobic groups just forget about hydrophobic groups from other protein molecules and associate rather with the additive. Such added substances may be vegetable oils. These oils contain hydrophobic and hydrophilic groups. The hydrophobic groups of the oil combine with the hydrophobic groups of the protein so that the latter can no longer associate with other hydrophobic protein groups. The hydrophilic groups of the oil, on the other hand, like the water and maintain the solubility of the protein molecules (Fig. 27).

Egg-white coagulation is a very temperature dependent process as all mountain climbers know. Eggs cannot be cooked in the shell to hard-boiled eggs in open containers on high mountains. The barometric pressure decreases rapidly with increasing altitude and water thus boils at lower temperatures which are not high enough to coagulate egg white. However, the higher pressure of a pressure cooker causes a higher boiling temperature of the water and denaturation will occur again.

8.6 TV Dinners and Stale Bread

The denaturation of protein molecules is a helix/coil transition with subsequent intermolecular association of the coil segments. The reverse process, coil-to-helix transformation, is not always desired. For example, regular corn flour can not be used as thickener for deep-frozen, ready-to-use sauces and gravies. Flours consist mainly of polymeric carbohydrates (see Table 18) and these mainly of starch. The unbranched amylose molecules of starch are imbedded in a continuous network of branched amylopectin molecules. This network prevents the formation of helices. On dissolution of starch in hot water, amylose molecules dissolve as more or less random coils. Their preferred structure is however not the coil but the helix. Relieved from the constraints of the amylopectin, they slowly form helical structures. The helices then assemble laterally and crystallize.

Such a "retrogradation" occurs also with common corn starch as thickener. These starches contain relatively high amounts of amylose (Table 9) which slowly forms helices. The texture and the appearance of these deep-frozen foods is not very pleasant; they thus do not sell very well. Such foods contain therefore amylose-poor (i.e., amylopectin-rich) starches from corn, barley or rice as thickener.

The staling of bread is probably also caused by an unwanted coil/helix transition. Bread doughs are high starch foods; their amyloses slowly retrograde into the helix conformation. Helices are however quite introverted fellows; their binding groups are rather busy with themselves instead of contacting the chemical sensors of our taste buds. They therefore create another taste then the coil-like amylose molecules, just that what we call a stale taste.

Staling depends on the temperature, however. Helices are rapidly formed at higher temperatures. But before the helix formation is complete, helical segments are grabbed by helical segments of other amylose molecules. The helical segments remain incomplete and the resulting physical cross-linking creates a highly swollen network; the amylose gelates.

The helix formation and the mobility of amylose segments is very slow at low temperatures. A temperature must thus exist at which well-formed helical segments are formed relatively fast without incorporation into intermolecular associations. For doughs, this temperature is just above the freezing point of water. Moral: Bread should never be stored in the refrigerator since it will stale fastest at its temperature. Room temperature yes, freezer yes, but never in the refrigerator.

The staling is slower if the flour still contains the gluten of the grain. Gluten is a protein which binds its peptide groups via hydrogen bonds to the hydroxy groups of the glucose, similar as vegetable oils bind to proteins (Fig. 27). This binding prevents the helix formation of amylose. Whole-grain breads contain much more gluten than white flours; their staling is much slower.

9 Spiders, Weavers, and Webs

"And I haven often thought that there might be a way found out, to make an artificial glutinous composition, much resembling, if not full as good, nay better, than that Excrement, or whatever other substance it be out of which, the Silk-worm wire-draws his clew. If such a composition were found it were certainly an easie matter to find very quick ways of drawing it out into small wires for use".

Robert Hooke, "Micrographia, or Some Physiological Descriptions of Minute Bodies Made by Magnifying Glasses with Observations and Inquiries Thereupon", Royal Society, London 1665

9.1 Silk, Artificial Silk, and Polyester

No wonder that Robert Hooke was so fascinated by silk and not by wool, cotton, linen, hemp or other natural textile fibers known to him. Silk was precious and expensive; it formed the clothes of the privileged. A gentleman's dress of dyed silk cost between 20 and 45 Florentine gold florins (guilders) in the middle of the sixteenth century, equivalent to the cost of 5 to 9 cows. Silk was beautiful to dye and was soft on the skin. Plebs had to wear garments of coarse linen from flax or of itching wool from sheep. No wonder that natural scientist Hooke (1635–1703) wanted to see an artificial silk.

Silks, or more precisely, natural silks, are produced by a number of caterpillars and spiders. The silk threads produced by certain spiders are so thin and still so strong that they were the only materials suitable for the cross wires of telescopes and binoculars not too long ago.

The most important silk comes from the mulberry silkworm (*Bombyx mori* Linné). It is sold as "mummies" but this has nothing to do with Egyptian mummies. The word is based on the Japanese "momme", a unit of weight (3.75 g), and denotes the fabric weight per square meter. Most silk is 16–22 mummies but certain Chinese silks are only 4–8 mummies.

The silkworm spins a cocoon which consists of 78% silk fibroin (the silk fiber) and 22% sericin (the silk glue). The pupa in the cocoons are killed by steam or hot air. The sericin is softened by dipping the cocoons into hot water so that rotating brushes can grab the ends of the silk threads of each cocoon. Four to ten threads are wound together on a winch and dried. Only 900 meters of thread can be used of the 3000–4000 meters in a cocoon; the outer and inner layers are not clean enough and are processed further together with damaged cocoons in a so-called schappe mill. No wonder that such a technology has caused the annual world production of silk to stagnate at 55000 tons for many years. No wonder also that silk is used almost exclusively for ties and scarfs in Western countries and that exquisite kimonos command prices of $30000 and more in Japan.

Legend has it that silk was already known to the Chinese emperor Huang-Ti in 2600 B.C. The secrets of silk manufacture were strongly guarded by the various rulers but in 555 A.D. Christian monks managed to smuggle silkworm eggs out of China, allegedly in a hollow cane. Sicily became a center of silk weaving after being conquered by Arab armies in 827. The silk industry of Lyon, France, was established during the sixteenth century but in the nineteenth century, the existence of the French silk industry was threatened by a disease of the silkworm.

The pending economic disaster instigated a search for an artificial silk. Chardonnet silk from cellulose nitrate was the first artificial silk (see Chap. 2.2) but these fibers had the unpleasant habit to burst into flames near ovens. No wonder, since they were nothing but gun cotton.

The viscose fibers fared better. They were invented in 1892 by the Americans Charles F. Gross, Edward J. Bevan, and Clayton Beadle. These fibers are now traded as "rayon" because of their shiny appearance. In rayon production, cellulose of wood pulp is treated with caustic soda (a solution of sodium hydroxide, NaOH, in water) to give alkali cellulose

$$(9\text{-}1)$$

$$\text{cell–OH} \xrightarrow[-\text{H}_2\text{O}]{+\text{NaOH}} \text{cell–ONa} \xrightarrow{+\text{CS}_2} \text{cell–O–C}\diagup^{\text{S}}_{\diagdown\text{SNa}} \xrightarrow[-\text{CS}_2;\,-\text{NaHSO}_4]{+\text{H}_2\text{SO}_4} \text{cell–OH}$$

cellulose	alkali cellulose	cellulose xanthogenate	cellulose

The alkali cellulose is then reacted with carbon disulfide CS_2 to give the orange cellulose xanthogenate (Greek: xanthos = yellow). The caustic soda solution of cellulose xanthogenate is viscous and hence called "viscose" (Latin: viscus = bird glue from the berries of the mistle toe). Viscose is spun through spinnerets into a precipitating bath of sulfuric acid whereupon the cellulose xanthogenate is re-converted into cellulose which is insoluble in sulfuric acid and immediately forms fibers. Rayon fibers have a silky appearance and were thus called "artificial silks" but this name is no longer used for fibers from cellulose and cellulose esters.

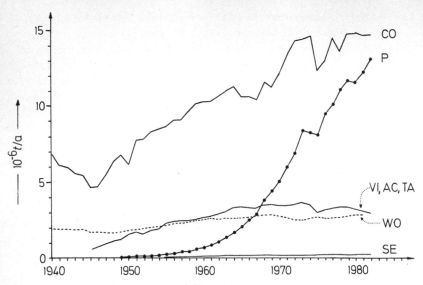

Fig. 28. Annual world production of fibers in tons per year. Cotton (CO), synthetic fibers (P), rayon and acetate fibers (VI, AC, TA), wool (WO), and natural silk (SE)

Rayon is a semi-synthetic fiber since its raw material cellulose is a natural product. Its world production has stagnated for many years (Fig. 28). This diagram does not disclose, however, that the rayon production has actually decreased in the Western countries. Not only are the fiber properties of rayon surpassed by those of many synthetic fibers (see below), but its manufacture also pollutes air and water. The high environmental protection costs make rayon production uneconomic in Western countries. Not so in the Eastern block countries and in developing countries where the rayon production *sans* environmental protection is still increasing. Textiles are more important than dirty rivers in such countries; priorities are not only a question of perceived or actual capitalist greed but also of the imagined or real necessities of life.

Rayon was followed by a great number of fully synthetic fibers (Table 19), one of which we know already as nylon. Several hundred synthetic fibers have been synthesized but only four groups of fibers have become important. Polyesters, polyamides, acrylics, and olefinics make up 98% of the world production of synthetic organic textile fibers.

Why does silk have such extraordinary properties? After all, sewing threads from silk and not from wool or cotton were used in the good old times. Well, a sewing thread must be tear resistant but the tear strength of wool is only 200 MPa whereas that of silk may reach 600 MPa (Table 20).

Table 19. Names, trade names, abbreviations, and chemical structures of important textile fibers. ETL = European Textile Law

Names			Abbreviations		Chemical structure of repeating unit
Common language	Chemical name	Trade names	Technical	ETL	
Wool	Mixture of various proteins	—	Wo	WO	$-NH-CHR-CO-$ with various R
Silk	Mixture of various proteins	—	Ms	SE	$-NH-CHR-CO-$ with various R
Nylon or polyamide	Polycaprolactam	Perlon, Grilon	PA 6	PA	$-NH-(CH_2)_5-CO-$
	Polyhexamethyleneadipamide	Nylon, Celon	PA 6.6	PA	a
Polyester	Polyethylene terephthalate	Dacron, Diolen, Fortrel, Grilen, Tergal, Terital, Terlenka, Terylen, Trevira	PE, PES, PET, PETP	PL	b

Acrylic	Polyacrylonitrile	Acrilan, Dralon, Orlon	PAC, PAN	PC	$-CH_2-CH-$
					        $\mid$
					       CN

Cotton	Cellulose	—	Bw	CO	
Linen	Cellulose	—	LI	LI	c
Half-linen	Cellulose	—	HF	HF	
Rayon	Cellulose	many	—	VI	
Triacetate	Cellulose triacetate	Arnel	CT	TA	like cellulose but with $O-CO-CH_3$ instead of OH
Acetate	Cellulose 2½ acetate	Acetate	CA	AC	like cellulose triacetate but with 2½ acetate groups per glucose residue instead of 3

a $-NH-(CH_2)_6-NH-CO-(CH_2)_4-CO-$

b $-O-CH_2-CH_2-O-OC-\langle\!\langle\,\rangle\!\rangle-CO-$

c

Cotton is about as tear resistant as silk but cannot be stretched that much (maximum elongation of 8% vs. 25%). A sewing thread must be "elastic", however, that is stretchable, since strongly varying pulling forces are caused by hand sewing but the seam should be very regular.

Table 20. Properties of some fibers. Resistance is indicated by 6 = excellent, 4 = good, 2 = satisfactory, and 0 = not satisfactory

Fibers	Modulus of elasticity in GPa	Tensile strength in MPa	Elongation in % (dry)	Resistance against		
				Abrasion	Alkali	Acids
Silk-like						
Natural silk	7–10	350–600	20–25	4	0	0
Nylon 6.6	6	800	19	6	5	0
Polyester	15	900	20	5	3	4
Kevlar	150	2800	4	5	4	2
Wool-like						
Sheep's wool	1– 3	150–200	30–40	2	0	0
Acrylics	5	280	35	3	2	4
Cotton-like						
Cotton	6–11	250–800	6– 8	2	6	0
Rayon	9	340	25	2	6	0
Hemp	29	850	2	2	6	0

The mechanical properties of threads are measured by so-called stress/strain experiments in which a fiber is slowly pulled apart. The resulting stress is then plotted against the elongation (Fig. 29).

Fibers respond differently to these experiments. Small elongations require large stresses in silk. Wool, on the other hand, stretches more and more by only small increases in the stress. Wool tears if it is elongated by more than 40%; this value is called the "tenacity" of wool. The stress at this elongation is the tensile strength at break. It is measured for almost all materials as area-related force (stress), for example, in $MPa = N/mm^2$. Only in textile science is this value given traditionally (and incorrectly) as "specific tensile strength" in g/tex where 1 tex = mass in grams of a fiber of 1000 meter length. Instead of the tex, the denier (1 g/900 m) was and is still used.

The stress does not increase directly proportional to the elongation in fibers (Fig. 29). No fiber thus behaves as a Hookean body since such a body would be represented by a straight line in Fig. 29. The initial slopes of the lines in stress/strain diagrams can however be treated as if they originate from Hookean bodies; these slopes give then the modulus of elasticity of the fiber. The greater the initial slope, the higher is the modulus, and the stiffer is the fiber. Silk is thus a stiffer fiber than wool.

But why is silk such a stiff fiber compared to wool? Both silk and wool are proteins, macromolecular chains composed of various alpha-amino acid units $\sim NH - CHR - CO \sim$. Silk has a much more simpler structure

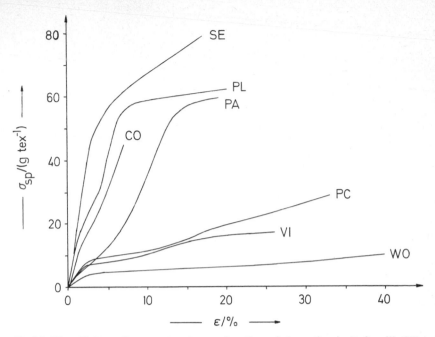

Fig. 29. "Specific" tensile stress in g/tex as function of elongation in % for silk (SE), polyester (PL), cotton (CO), polyamide 6 (PA), acrylics (PC), rayon (VI), and wool (WO)

than wool. The protein chains of the silk from *Bombyx mori* L. consist of a regularly structured part with ten segments of six amino acid residues each

H (ser-gly-ala-gly-ala-gly)₁₀OH

and an irregularly structured part of 33 amino acid residues. The regular chain segments can arrange themselves laterally (Fig. 30), similar to the chains in Fig. 27 z. This dense, regular packing of the chains is responsible for the high strength of silk in fiber direction. Any attempt to stretch the thread in fiber direction must first deform and cleave the intermolecular hydrogen bonds, then deform the bond angles between the chain atoms and finally stretch the bond lengths. All this has to occur simultaneously for many bonds which requires lots of energy.

The situation is much more simple for the other 33 α-amino acid residues. These residues are chemically very different and they thus do not pack easily in a regular manner. Their chains rather adopt the overall conformation of a random coil. Such coils are easy to deform; these amino acid segments are responsible for the high extensibility of silk.

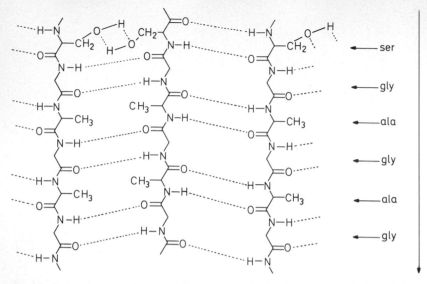

Fig. 30. Beta structures of the crystalline regions of natural silk, so-called pleated sheet structures, since the structures are folded like pleated sheets along the chain directions in three-dimensional space. The large arrow indicates the fiber direction, the dotted lines give the hydrogen bonds between amide groups

The highly ordered parts of silk fibers are called "crystalline" although they really do not look like nice quartz crystals. Not only are they so small that they cannot be seen by the naked eye (nor under a microscope!), they also do not show any of the characteristics of a classic crystal: solid materials whose plane surfaces intersect at defined angles. This nineteenth century definition of a crystal has been broadened in modern science. No longer is the external shape (the habitus) the important property of a crystal. It is rather the inner structure, that is the regular arrangement of atoms and atomic groups, which now defines a crystal. Such structures may be probed by x-rays but they cannot always be seen by microscopes or electron microscopes. Fig. 31 shows, for example, four different morphological structures of the same polyamide 6. Structure a) is not crystalline, but structures b)–d) are. Structure d is even 100% crystalline according to x-ray measurements. Whether or not chains pack to give crystalline regions in polymeric materials will mainly depend on their overall conformation. Helices and chains in all-trans conformations possess ordered structures; they arrange themselves easily in crystalline regions. Coils are not ordered; they often pack in solids as non-crystalline, "amorphous" materials (Greek: a = none, morphe = shape, form).

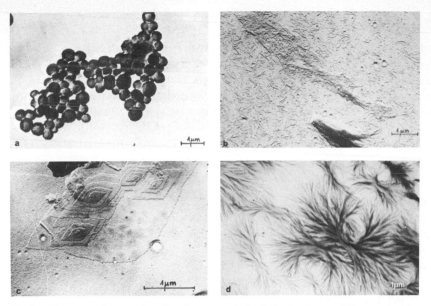

Fig. 31. Electron micrographs of various morphological structures of the same polyamide 6 (after Ch. Ruscher and E. Schulz). The pouring of a 260 °C glycerol solution of PA 6 into cold glycerol gives spherical, amorphous structures (a). The slow cooling of the same glycerol solution with 1–2 °C/min results in fine needles (b), the fast cooling in crystal-like lamellae (c). The slow drying of a PA 6 solution in formic acid gives dendritic structures (d). (b)–(d) are crystalline but to different extents

Crystalline regions stiffen polymeric materials and reduce their extensibility. The same polymer can thus give materials with very different properties depending on its crystallinity. In low molar mass materials, however, the chemical structure creates in general only one physical structure which in turn dictates the macroscopic physical properties. Examples are cane sugar, mercury, or nitrogen. In polymers, on the other hand, various physical structures and thus properties may be created by different processing methods. Temperatures, flow rates, and additives may all influence the degree of crystallinity and sometimes also the crystal structure (see also Fig. 31). In polyamide 6 (nylon 6), $H(NH-(CH_2)_5-CO)_nOH$, for example, a low crystallinity of less than 15% is desired for shopping bags, but a high one of over 90% for fishing lines. The crystallinity of nylon 6 for ladies' underwear should be about 20–30%, for hosiery 60–65%, and for tire cord 75–90%.

Natural fibers are still the yard sticks with which all textile fibers are measured. Wool is warm and shape resistant, cotton is water absorbing and easy to wash, and silk is cool and elegant. Many of these properties can be traced back directly to the chemical structure of macromolecules, for example, the water uptake of cotton is related to the hydrogen-bonding properties of the hydroxyl groups of cellulose. Other properties derive from the physical structure of polymers; for example, the high strengths of silk and polyamides from their crystallinities. Still other properties depend on the structure and shape of the fibers themselves.

Silk consists of smooth, "endless" fibers of maximal 4000 m length (Fig. 32); such endless fibers are called filament fibers. The triangular cross-section of silk fibers reflects light in all directions, similar to the cut surfaces of diamonds, and cause the glossy shine of silk. The triangular cross-section and the smooth surface are also responsible for the feel ("hand") and the rustling noise of silk.

Wool exhibits short fibers of maximal 13 cm length (Figs. 32 and 33). Wool is not a filament fiber like silk but a staple fiber as such short fibers are called. The wool fibers are more or less round (Fig. 33). Their scaly sur-

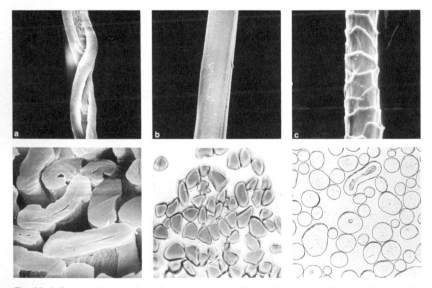

Fig. 32. Microscopic pictures of various natural fibers. Upper row: fibers of cotton (a), silk (b), and sheep's wool (c). Lower row: corresponding cross-sections

face creates the typical hand of wool. The crimping and the bulky volume, on the other hand, are created by the bicomponent structure of wool fibers. The two halves of a wool fiber, para-cortex and ortho-cortex, are chemically slightly different and absorb different amounts of water. One half swells more than the other and the wool fiber bends and crimps.

Similar properties can be given to modern synthetic fibers, just by changing the fiber structure. Circular holes in spinnerettes create fibers with circular cross-sections, but trilobal holes lead to triangular ones and tripropellers to trilobals (Fig. 34).

The reasons for these changes in cross-sectional patterns are not very difficult to understand. Polymers are spun either from the melt or from concentrated solutions. In both cases, most polymer molecules are present as random coils. If these coil molecules are forced through a hole, as in a spinnerette, coils are squeezed vertically to the flow direction and extended in flow direction itself (Fig. 35). On leaving the hole, coils try to revert to their unperturbed dimensions. They expand most where they were squeezed most, for example, near the center of a propeller. Bicomponent fibers with a core/cover structure result if fibers are spun from two different polymers through two concentric holes (Fig. 36). Spinning through two different halves of the hole gives bicomponent fibers with side-by-side structures. If fibers are dispersed in polymer melts or solutions, matrix/fibril structures are created. The number of modifications by textile technology is almost limitless. No wonder that modern textile science can prepare synthetic fibers with almost any desired property:

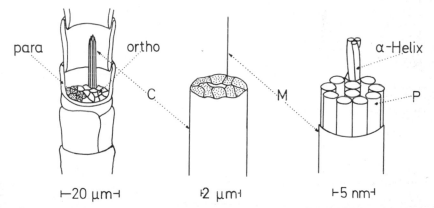

Fig. 33. Structure of a wool fiber with para-cortex and ortho-cortex (left). The cortices consist of bundles of cortex cells (center). Each cortex cell contains microfibrils (right) which in turn are composed of 11 protofibrils. Each protofibril consists of 2 to 3 α-helices of protein chains

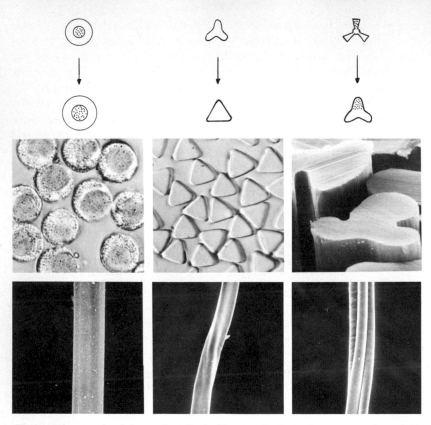

Fig. 34. Microscopic pictures of synthetic fibers and schematic representation of the spinneret holes used to manufacture the fibers. From top to bottom: spinneret holes, resulting cross-sections of spun fibers, actual fiber cross-sections, fibers. From left to right: core/cover bicomponent fiber from polyamide 6 and polyamide 6.6, polyester fiber, bicomponent acrylic fiber with trilobal cross-section

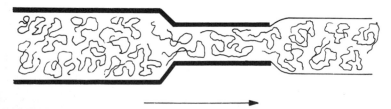

Fig. 35. Deformation of random coils on flowing through a nozzle and subsequent coil expansion after leaving the nozzle

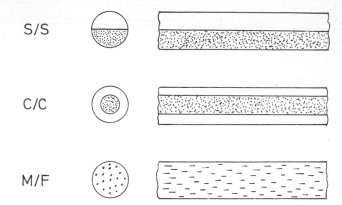

Fig. 36. Bicomponent fibers. S/S = side-by-side, C/C = core/cover, M/F = matrix/fibril

Fibers from polyacrylonitrile with high crimp for wool-like fibers, polyamide fibers with high strengths for silk-like fibers, polyester fabrics with good shape retention for suits, water absorbing fibers for active sportswear, elastic fibers, waterrepelling, stiff, soft, flame-retardent, warming, cooling, etc.

How about a suit which is warm and wool-like in winter but cool and silk-like in summer? The perfect dress for the globetrotting executive! Possible? Well, in principle yes but ... Honestly, it would almost be possible. The poly(α-amino acid) polyleucine

$$\sim\!\!\sim (NH-CH-CO) \sim\!\!\sim$$
$$| \atop CH_2-CH(CH_3)_2$$

adopts a helical conformation in certain solvents and may be spun into wool-like fibers. Treatment of these fibers with other solvents transforms the helices into pleated sheet structures (similar as in Fig. 30) and the fibers become silk-like. The conversion back into wool-like fibers is a bit troublesome, though ...

And now the hundred thousand dollar question: How about the energy? Aren't synthetic fibers too costly with respect to the precious petroleum, the polluting coal, the nature destroying hydroelectric power stations, or the absolutely unnecessary atomic reactors? Why not stay with good ole cotton which needs only sunlight and water? By the way, it is only a rumor that first-class cotton must be harvested by sacred virgins at new moon. Well, sacred virgins are not really necessary, but the new moon may indeed be beneficial. Cotton is the seed hair of a bush. If the seed cap-

Table 21. Energy requirements for the manufacture and care of men's shirts

	Physical unit	Energy requirement		
		100% cotton	65% polyester; 35% cotton	100% polyester
Life cycle of a shirt (number of washes)	–	50	75	75
Energy requirements for the manufacture of				
the fiber	MJ/kg	49	–	179
the shirt	MJ/shirt	95	117	78
Energy per wash	MJ/wash	1.9	1.6	1.0
Energy for textile care, modern amenities				
Automatic washer	MJ/wash	2.3	1.1	1.1
Automatic dryer	MJ/wash	2.9	1.3	0.6
Electrical ironing	MJ/wash	1.2	0.4	0
Total per wash	MJ/wash	6.4	2.8	1.7
Energy for textile care, traditional households (hand wash, air dry)				
Total per wash	MJ/wash	0.6	0.3	0.2
Total energy consumption for manufacture and care				
Modern household	MJ/wash	8.3	4.4	2.7
Traditional household	MJ/wash	2.5	1.9	1.2

sules are opened too early in full sunshine, light and oxygen of air start a photochemical oxidation of the cellulose chains of cotton. The long chains are degraded to shorter ones and the mechanical properties of cotton are reduced. This influence of the degree of polymerization (or molar mass) on the properties can be easily seen: the high molar mass cellulose, a polymer of glucose with degrees of polymerization of up to 14000, is a good fiber but cotton candy from cane sugar (degree of polymerization 2) does not make good fabrics.

Back to the energy problem: aren't natural fibers more advantageous than synthetic fibers? Well, cotton does not grow "from alone", at least, if one wants to produce it economically. Energy is needed to make fertilizers, run cotton-picking machines, and operate irrigation systems. Cotton does need energy for its manufacture. Still, the energy required to grow cotton fibers is much lower than the energy needed to manufacture polyester fibers (Table 21). But this is not the whole story. The cotton fibers must be prepared for textile use and the energy consumption for the manufacture of cotton shirts is already higher than the one for polyester shirts! Shirts from polyester last longer than those from cotton since polyester is more abrasion resistant (see also Table 20). Cotton shirts are more expensive to care for, whether this is done with the wash-board in the river or by pushing a button at the washing machine. In any case, the cost per wash is twice or triple of that for polyester and the total cost of a cotton shirt over its lifetime is much greater than that of a polyester or blended shirt as shown in Table 21.

9.3 Paper and Leather

Paper is a sheet-like porous product, already known to the Chinese around 200 B.C.. Classic papers are felts of short fibers; they are fleeces. New synthetic papers are often plastic films. Synthetic papers are presently not much consumed, however. More than 99% of all papers are cellulose papers, and of these 95% are made from wood, the rest from straw, cotton, linen, bamboo, and bagasse, the dry pulp from sugar cane. The papyrus plant, which gave the name to paper, is no longer used for paper making.

The world uses annually about 140 million tons of paper. Since green wood contains about 40–60% water and the wood solids about 45% cellulose (Table 22), paper consumes about 600 million tons of wood per year. Furthermore, about 800 million tons of wood are cut each year for lumber and firewood. The total wood on earth is estimated to be ten thousand million tons. The annual consumption is thus 1.4%. No reason to be concerned: if trees grow to be 70 years old, about that many are cut as grow anew.

Table 22. Solids content of some trees and plants

Name	Content in %			
	Cellulose	Lignin	Hemicelluloses	Ash
Bamboo	60	24	14	2
Bagasse	44	24	29	3
Pine	43	29	27.7	0.3
Poplar	43	23	32.3	0.7
Straw (wheat)	40	17	36	7

Table 23. Forested areas

Continent or country	Areas in million square kilometers				Portion in %	
	Total area	Forested area	Accessible forest	Utilized forest	Forest as % of total land	Utilized as % of accessible
Asia[a]	42.2	5.0	3.1	2.3	12	75
Africa	30.3	7.0	2.8	1.1	23	38
Soviet Union	21.9	7.4	4.3	3.5	34	82
Central and South America	19.7	11.0	3.3	0.83	56	25
North America	19.1	7.1	3.1	2.2	37	71
Europe[a]	10.6	1.40	1.33	1.30	13	98
Australia	7.9	0.92	0.20	0.17	12	85
Greenland	2.2	0	0	0	0	0
Total	153.9	39.8	18.1	11.4	26	63

[a] Without Soviet Union.

But these are statistics and statistics predict on average a cozy warmth if one foot is in boiling water and the other one on ice. Europe uses 98% of the accessible forests (North America 71%) but forests cover only 13% of the total land area of Europe (North America: 37%) (Table 23). European countries are thus much more concerned about wood consumption and forests dying from acid rain than other countries.

About 50% of the paper is used for graphic papers (newsprint etc.), 40% for packaging, and 7% for hygienic purposes. Should one not save trees and replace packaging papers from cellulose by those from synthetics? How about recycling? Or incineration? Should one not use less packaging even if more foodstuffs spoil and the whole distribution system has to be changed? These questions do not call for simplistic answers and it is the consumer who pays anyway.

Coniferous trees are used for 75% and deciduous trees for 25% of the cellulose papers manufactured in the United States and in Europe. Bark is removed and the wood is then cleaned, cut into pieces and ground. Cellulose fibers are separated from the other wood components, washed, bleached, cleaned, and mixed with dyestuffs, resins, sizes, and other additives. The resulting aqueous dispersion runs through a brass sieve in the paper machine; water and fine particles drop off and a wet fleece is formed. The fleece is transported through roll presses and cylindrical dryers; the residual water evaporates and the cellulose fibers bond to each other.

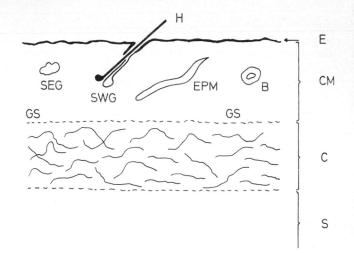

Fig. 37. Structure of the skin (schematic). E = epidermis, CM = corium minor, C = corium, S = subcutis, GS = ground substance, SEG = fat glands, SWG = sweat glands, H = hair, EPM = erector muscles, B = blood vessels

Fibers should lay parallel to the surface in papers. In true fleeces, short fibers should be oriented perpendicular to the surface, however. The fibers of fleeces are then bonded, for example, mechanically by needling, physico-chemically by glueing, or thermally by welding. Such fleeces serve for carpet backing or for inlays in suits.

Leather is also a kind of fleece of chemically or physically crosslinked collagen fibers from animal hides and skins. Hides consists of various layers. The thin upper layer (epidermis) is the carrier of the biological skin functions. It is followed by the corium minor where blood vessels, sweat glands, etc., are embedded in a so-called ground substance (Fig. 37). Below the corium minor is the corium and the subcutis which connects the skin to the muscles.

Epidermis and subcutis are removed in leather making. Only the corium, a thick fleece of collagen fibers, remains. Collagen is a right-winding superhelix consisting of three left-winding peptide chains with more than 1000 α-amino acid residues each.

The tanning of leather crosslinks the collagen fibers by reaction with multifunctional reagents under preservation of the fiber structure. Depending on the origin of the hides (young or old, calves, cows, goats, sheep, etc.), the crosslinking agents (amino resins, chromium compounds, etc.), and the other additives (plasticizers, repellants, etc.), various types of leather can be made.

Scarcities and the labor intensive preparation of natural leathers stimulated the search for synthetic leather-like materials decades ago. The long known artificial leathers simulate the structure of natural leather by coating fabrics with plastics. These artificial leathers serve well for handbags and luggage but not for the upper shells of shoes. The plastics used (polyvinyl chloride, polyamide, polyurethane) are rather rigid so that they do not fit the feet that well. They are also not porous and do not "breathe".

The problem of breathing was solved by the development of so-called poromerics, artificial leathers which are permeable for water vapor. Poromerics consist of polyester or polyamide fabrics which are impregnated by polyurethane or synthetic rubbers and coated with one or two layers of polyurethanes.

Totally different are the synthetic suedes (Ultrasuede in the USA, Alcantara in Europe, Ecsaine in Japan, Australia, and South Africa). Extremely thin polyester fibers are embedded in polystyrene to form matrix/fibril fibers (see Fig. 36). These M/F fibers serve to make a needle fleece which is subsequently impregnated by polyurethanes. The polystyrene is extracted by the solvent dimethylformamide. Freely moving polyester fibers remain in pores of a polyurethane network, similar to the collagen fibers in the corium (Fig. 38). This synthetic suede is "lighter" and more wrinkle-resistant than natural suede; it is also washable.

10 How to Iron Correctly

> *The average polymer*
> *Enjoys a glassy state, but cools, forgets*
> *To slump, and clouds in closely patterned minuets.*
>
> John Updike, The Dance of the Solids, in
> "Midpoint and Other Poems", A. Knopf, New
> York 1968

10.1 Glasses and Glass Transitions

Wet cotton cloth, rubber tires, plastic table covers from polyvinyl chloride, glass melts, and lathered beards, what do they have in common? Why does chewing gum becomes soft in the mouth?

To borrow from poet Updike: all these materials are average polymers which would rather enjoy the glassy state. The "closely patterned minuets" testify however to the freedom of speech and creativity of the poet. It is just the opposite, if a minuet at all, then above the glassy state, not in it. Molecules move very little in the glassy state and certainly not in graceful dances. Despite the poet's assertion, many polymers do not cloud if and when they reach the glassy state but stay glass clear. Only: what is a glassy state anyway?

Cotton, rubber, polyvinyl chloride, hair and chewing gum are macromolecular compounds. Their chain atoms can rotate more or less freely around the chain bonds (Chap. 7) provided their environment lets them do this. Chains and chain segments are separated from each other in dilute solution by many solvent molecules; the rotation is not hindered by the presence of other chain segments. On the other hand, chain segments are packed so tightly in the crystalline state that chain atoms can move to a minor degree only.

How is it in non-crystalline (amorphous) solid materials? Or is this a wrong question: aren't all solid materials crystalline? Quartz, iron, and cane sugar are solid crystalline materials at room temperature. Why are some solid materials non-crystalline?

Iron is made up by spherical iron atoms. Heating delivers energy to the atoms; they race around and do not stay long at any position. The energy needed for these movements is missing at temperatures below the melting temperature; each atom tries to occupy a quiet site not too far away from its neighbors since nature abhors voids. Look into a pot with equal-sized marbles: they are tightly packed with high order, that is, they form a macroscopic (crystalline) lattice. Iron atoms do it in the same way.

Life is not that simple for cane sugar. Saccharose (cane sugar, beet sugar, sucrose) does not consist of spherical molecules, but slightly elongated ones (see Chap. 6.2). Such molecules have more difficulty in order to arrange themselves in regular arrays. Their hydroxyl groups also love to form hydrogen bonds with other hydroxyl groups without being too discriminating where these other groups are coming from. Honey, a concentrated aqueous solution of the sugars saccharose, glucose and fructose, crystallizes only after long times because of such difficulties.

Crystallization is even more difficult for chain molecules. The polymer chains are inherently anisotropic since the chain length is much greater than the chain diameter (see Chap. 3.5). Chain molecules are not easy to motivate to arrange themselves laterally but a little friendly force will sometimes work wonders. Matches or uncooked spaghetti are good examples. Chain substituents should not disturb the crystalline order, so they must be arranged regularly along the chain: isotactic and syndiotactic polypropylene (Fig. 21) crystallize but not atactic polypropylene.

On cooling a polymer melt, chain molecules cannot immediately find their ideal crystalline positions, if at all. Non-crystalline or partly crystalline polymers result, the latter with different morphologies, depending on chain structure and experimental conditions (Fig. 31).

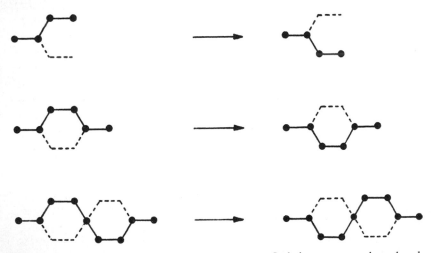

Fig. 38. Some cooperative motions of chain segments. ● chain atom, —— valence bonds before the motion, --- valence bonds after the motion

Chain segments are less tightly packed in the amorphous regions and thus chain atoms have more freedom if they want to occupy a new position by rotation around a chain bond (Fig. 13). These chain atoms are however bound left and right to other chain atoms and if they want to rotate, other chain atoms have to follow, whether they want to or not (Fig. 38). Adjacent chain segments also have to move to make space. The rotation around a chain bond thus requires coupled (cooperative) intramolecular and intermolecular movements of other chain segments. Try to move one spaghetti (our chain molecule) in a pile of other spaghetti (our solid polymer). It can not be done without moving other spaghettis.

Changes of spatial positions are thus not easy in macromolecular compounds. Cooperation of adjacent chain segments of the own chain and cooperation of neighboring chains is required. The cooperative movement is only possible if a certain energy barrier is overcome, that is, at a certain temperature called the glass transition temperature or glass temperature.

This characteristic temperature got its name because it was first observed with silicate glasses, the normal glass of our windows. Silicate glass consists at room temperature of partially branched silicate chains

$$\begin{array}{c}
\overset{O^{\ominus}}{|} \quad \overset{O^{\ominus}}{|} \quad \overset{O^{\ominus}}{|} \quad \overset{O^{\ominus}}{|} \quad \overset{O^{\ominus}}{|} \quad \overset{O^{\ominus}}{|} \quad \overset{O^{\ominus}}{|} \quad \overset{O^{\ominus}}{|} \quad \overset{O^{\ominus}}{|} \quad \overset{O^{\ominus}}{|} \quad \overset{O^{\ominus}}{|} \quad \overset{O^{\ominus}}{|}\\
\text{---} O - Si - O - Si - O - Si - O - Si - O - Si - O - Si - O - Si - O - Si - O - Si - O - Si - O - Si - O - Si \text{---}
\end{array}$$

The branches prevent the silicate chains to arrange themselves in crystalline lattices. The thermal energy provided at room temperature is not sufficient to allow chain segments to move (at least not at short times) and the silicate glass appears as a solid material. Silicate chains and chain segments are not totally motionless at room temperature, however: if a glass rod is supported at both ends, its center will very slowly bend under its own weight.

The thermal energy supplied at the glass temperature is sufficient to cause relatively rapid cooperative movements: the glass "melts". This process should not be called "melting", however, since this term is reserved for the transformation of a crystal into a liquid. The "melting" of glass is rather the proverbial glass transition, the transformation of a glassy, noncrystalline body into a highly viscous, "rubbery" material. Silicate glass is

121

Table 24. Glass transition temperatures of some polymers

Polymer	Glass temperature in °C		Melting temperature in °C
	dry	wet	
Silicones	− 123	−	−
Polyethylene	− 90	−	135
1,4-*cis*-polyisoprene (natural rubber)	− 68	−	36[a]
1,4-*trans*-polyisoprene (guttapercha)	− 60	−	70
Polypropylene, isotactic	− 15	−	176
Polyvinyl acetate	28	−	−
Polycaprolactam (perlon)	50	?	225
Polyethylene terephthalate (polyester)	69	?	270
Polyvinyl chloride	90	−	−
Silk	162	?	250
Cellulose (cotton, rayon)	225	20	?
Wool	?	< 100	?
E glass	840	−	−

[a] Crystallizes only under tension or at very low temperatures.

not crystalline whereas quartz, $(SiO_2)_n$, is. Quartz "glass" may or may not be crystalline.

All solid organic and inorganic macromolecular compounds exhibit glass transition temperatures if they contain amorphous regions. Only in these regions can segments move cooperatively. Partly crystalline polymers possess both crystalline and amorphous regions; they show correspondingly both melt and glass transitions. Completely crystalline polymers melt but do not show glass transitions.

Glass transition temperatures vary widely for different polymers since they depend on two properties, the stiffness of the chains themselves and the interactions of the chains with their neighbors. Stiffer chains and stronger interactions require more energy for cooperative movements; hence they increase the glass transition temperature.

Polydimethylsiloxanes ("silicones")[1], $\sim\sim(Si(CH_3)_2 - O)_n\sim\sim$, exhibit a very low glass transition temperature of − 123 °C (Table 24); they are viscous liquids down to that temperature. The glass transition temperature of chewing gums should be between room temperature (so that it behaves as a solid in the store) and the temperature of the mouth (so that it is soft on chewing). Chewing gum from chicle is based on a naturally occuring polymer, guttapercha, which is naturally plasticized by triterpenes. Most

---

[1] A more detailed treatment of the proportion of silicones is found in the excellent introduction "Silicon and Silicones" by E. Rochow (see Suggested Readings).

chewing gums are, however, based on polyvinyl acetate, a synthetic polymer with the structure $\sim\sim (CH_2 - CH(OOCCH_3))_n \sim\sim$.

Polyethylene has a very low glass transition temperature of $-90\,°C$, yet it is not a viscous liquid or a rubbery material at room temperature but a solid. The reason is that it is partially crystalline. The crystalline regions act as physical crosslinking agents between the polyethylene chains and polyethylene behaves as a solid. If the crystallinity of the polyethylene chains is destroyed, e. g., by copolymerization of ethylene with propylene, then the irregular methyl substituents of the propylene residues prevent the ordering of chains to form a crystal lattice. The copolymer from ethylene with propylene is amorphous and because of the low glass transition temperature of the polyethylene it is indeed a rubber. Such ethylene/ propylene copolymers are used, for example, as rubber membranes for flat roofs.

10.2 Ironing and Shaving

Cotton consists of cellulose molecules with a high glass transition temperature of $225\,°C$. The cellulose segments cannot move at room temperature; fabrics from cellulose fibers retain their shapes. Cellulose takes up water very easily, thus, it absorbs sweat and hence is very well suited for underwear and also for suits and dresses in tropical climates. The absorbed water resides between the cellulose chains and pushes them a little bit apart. Since water molecules are very small and can be easily moved, cooperative movements of cellulose chains are thus made easier and the glass transition temperature of cellulose "plasticized" by water is lowered to about $20\,°C$.

Because wet cellulose fabrics exhibit glass transition temperatures which are below the room temperature, that means that cellulose segments can move around, and under pressure the fabrics wrinkle. Ironing removes the wrinkles but it has to be done by putting a wet cloth between the fabric and the hot iron. The wet cloth insulates the fabric from the heat of the iron, in part by its very existence, and in part because water evaporates and withdraws energy and thus lowers the temperature. The water of the wet cloth also acts as additional plasticizer for the cellulose fabric so that the cellulose chains can move and be oriented more easily under the pressure of the iron, leaving a desired orientation of the cellulose chains and removing the wrinkles.

The melting temperature of cellulose is very high, and cellulose does not melt at regular ironing temperatures. Since cellulose would be degraded at these temperatures, cotton and rayon fabrics have to be protected against singing (degradation) by a wet cloth. The situation is quite different for fabrics made from polyamides (nylon, perlon, etc.). These polymers can melt and start flowing which alters the fiber and the fabric structure. The polar amide groups can also cause sticking to a hot iron. Nylon and perlon thus have to be ironed with a cooler iron than cotton and rayon.

The wet cloth can be replaced by a steam iron. The steam forms here the thermal protection layer for the fabric and the water acts again as plasticizer so that the chains can be easily oriented under pressure. High temperatures then remove the water from the fabric and the desired orientation of the chains is "frozen in". Smooth fabrics can also be achieved by tumblers. The steam acts again as plasticizer. The rotary movement of the tumbler causes the fabrics to float in space. Little or no pressure is exerted on the fabric, either by its own weight or by the weight of another fabric layer. The plasticized chain segments can thus easily attain their equilibrium positions which are then frozen in by the removal of water vapor and the lowering of the temperature. The fabrics become smooth without ironing, albeit not as much as with ironing. Light pressures always force the desired orientation of molecules – and men.

How about the ironing of wool? It is also done with a wet protecting cloth and a hot iron. Certainly, the same physical processes operate as in the ironing of cotton or nylon. In addition, a chemical process is present, however. The protein molecules of wool contain "disulfide bridges" –S–S– between different protein molecules and between segments of the same molecules. These disulfide bridges act as crosslinks and prevent the movements of chains and chain segments and are responsible for the high form stability of wool fabrics.

Disulfide bridges are broken and thiol groups –SH are formed when wool is treated with alkaline detergents. These thiol groups and the remaining disulfide bridges regroup on ironing, as shown in Eq.(10-1):

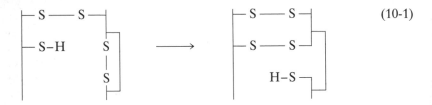

$$(10-1)$$

If wool is steamed short times, only bonds are broken but no new ones are formed. Reformation of disulfide bridges requires longer times and higher pressures. The ironing of wool thus needs heavier irons (or more pressure) and this is the reason why professional cleaners use laundry presses for woolen suits.

Exactly the same process is used by the hair dresser to make permanent waves (perms). Human hair is also a "wool", only slightly different from sheep's wool. Alkaline reducing agents are used to cleave disulfide bridges, for example, sodium hydrogen sulfite ($NaHSO_3$) in hot perms and ammonium thioglycolate ($HS - CH_2 - COONH_4$) in cold perms. The hair is then shaped by curlers and the new shape "fixed" by formation of new disulfide bridges with the help of oxidizing agents. The process also generates some low molar mass organic sulfur compounds which cause the characteristic smell of wool and hair.

10.3 Plasticizers

Water permeates into cellulose and protein fibers only slowly at room temperature. Human hair even exhibits an induction period during which no water is taken up at all. The water uptake is still slow at 40 °C, just above skin temperature (ca. 37 °C), but who wants to shave with boiling water just to plasticize the beard? All shaving creams thus contain surface active compounds which improve the wetting of the surface of hairs and ease the permeation of water into the hair. The beard is thus plasticized faster and at lower temperatures.

The plasticizing effect of solvents on polymers is not restricted to water as plasticizer and natural polymers like cellulose (cotton) and proteins (wool). As mentioned, chicle used in gum is a plasticized polymer. Plasticization was godfather at the birth of the first plastic, celluloid: camphor is a very good plasticizer for cellulose nitrate (see Chap. 2.3). Many plastics and most rubbers contain plasticizers.

Polyvinyl chloride (PVC) is a fairly rigid material which is difficult to process with conventional plastics machinery. PVC becomes very soft and easy to process after it is plasticized with low molar mass plasticizers like phthalic acid esters or citric acid esters. Plastic table clothes are often manufactured from plasticized PVC. Since these plasticizers are low molar mass molecules, little energy is needed for their evaporation. Such table clothes often release small amounts of plasticizer, especially initially and the table cloth smells like "plastic". But it is the plasticizer which smells, not the plastic.

The same unpleasant effect can sometimes be observed in cars. The interior side of car windows becomes covered with a sticky mass of plasticizer which evaporated from the plasticized polymers used in dashboards, car seats, etc.. Rear windows are especially vulnerable since they are near to the seats and the sun creates a severe greenhouse effect, often with temperatures up to 60 °C.

The evaporation of plasticizer from PVC can be reduced if plasticizers with higher molar mass are used, since these need more energy to convert to the gaseous state. Such plasticized PVC's are relatively trouble-free. Car designers can also use other materials. It need not to be leather, carpets for boards and fabrics for seats will do nicely.

Leather is also almost always plasticized. The washing of leather dissolves the plasticizer and the leather starts looking shabby. Such a change of appearance may also be caused by dry-cleaning which is not "dry" at all but uses organic liquids instead of water. Some of these liquids extract the plasticizer from the leather. Cleaners are often very reluctant to accept leather goods since they have no way to find out which plasticizers were used in its manufacture which is kept as a characteristic trade secret.

Plasticized polymers are not rare. Wood is a composite of cellulose and lignin which is plasticized with water (see Chap. 11.3). Tires are plasticized with large amounts of mineral oil but we will discuss this in Chap. 12.3.

11 From Cheap Substitutes to High Performance Materials

> *Polyethylene is good for inert laboratory beakers and very little else.*
>
> **R. E. Dickerson and I. Geis, *The Structure and Action of Proteins*, Harper and Row, New York 1969, page 4**

11.1 Thermoplastics and Thermosets

Authors Dickerson and Geis must be a little inexperienced in the matters of the world to make such a statement in their otherwise very nice book. Polyethylene is annually produced world-wide in over 15 million tons. A laboratory beaker weighs about 50 g and since such beakers can practically be used only for water solutions, their main consumers are biochemists and molecular biologists. The number of this species can be estimated as about 100 000. Since polyethylene is allegedly good only for laboratory beakers, each bio person has to use 3 000 000 beakers per year. Scientists are supposed to work 6-day weeks with 10-hour days, each scientist will thus have to consume 17 beakers per minute, which requires some artistic talent.

Fun aside: polyethylene is used only in negligible amounts for laboratory beakers. The major portion is used for packaging films, containers, pipes, and tubes. Tear-resistant polyethylene films cover meat trays in supermarkets and early crops in gardening. Polyethylene bottles serve for milk, paints, and pills. Polyethylene is also used for garbage bags, draining pipes, and as sheathing for electrical cables. The many other plastics find a multitude of applications; some typical ones are shown in Table 25.

In Western Germany in 1982, 25% of the plastics were used in construction/buildings, 21% in packaging, 15% in the electrical/electronics industry, 10% for paints and adhesives, 7% in transportation, 5% in furniture, 4% in agriculture, 2,5% for appliances and housewares, and 10.5% for other applications. Similar numbers apply to the United States and other highly industrialized countries for which only incomplete statistics exist (about 40% of the U.S. plastics use is unaccounted for!).

The ubiquitous plastics have become synonymous with "unnatural" as the terms "plastic society", "plastic money" etc testify. But are they really

Table 25. Important polymers from polycondensation (PC), ring-opening polymerization (RO), or addition polymerization of monomers with carbon/carbon double bonds (CC) or carbon/oxygen double bonds (CO)

Symbol	Polymer name	Monomeric unit	Synthesis	Typical applications
Thermoplastics				
PE	polyethylene	$-CH_2-CH_2-$	CC	Packaging films, bottles, cable sheathing

PP	polypropylene	$-CH_2-CH-$
		       |
		      CH_3
			CC	Car parts, irrigation pipes, outdoor carpets

PS	polystyrene	$-CH_2-CH-$
		       |
		      C_6H_5
			CC	Beakers, TV consoles, toys, foams for heat insulation and packaging

PVC	polyvinyl chloride	$-CH_2-CH-$
		       |
		      Cl
			CC	Pipes, tubes, artificial leather, shopping bags

PVDC	polyvinylidene chloride	$-CH_2-CCl_2-$	CC	Aroma barrier packaging films, non-burning doll hair

PMMA	polymethyl methacrylate	$-CH_2-C(CH_3)-$
		          |
		       $COOCH_3$
			CC	Organic glass for lamps, signs, and bullet-proof windows; sanitary cells

PTFE	polytetrafluoroethylene	$-CF_2-CF_2-$	CC	Electrical insulations, coatings for frying pans
POM	polyoxymethylene	$-O-CH_2-$	RO, CO	Precision parts, e.g., for toy trains
PET	polyethylene terephthalate	$-O-(CH_2)_2-O-CO-C_6H_4-CO-$	PC	Fibers, bottles

Table 25 (continued)

Symbol	Polymer name	Monomeric unit	Synthesis	Typical applications
PA 6	polyamide 6 (nylon 6)	$-NH-(CH_2)_5-CO-$	RO	Fibers, gears
Thermosets				
PUR	polyurethane	$-R-NH-COO-R'-OOC-NH-$	PC	Car parts, elastomers, fibers, foams
SIR	silicones	$-O-Si(CH_3)_2-$	RO, PC	Elastomers, implantates, water-repelling coatings
UP	unsaturated polyesters	$-O-(CH_2)_2-O-CO-\overset{\mid}{C}H-CH-CO-$	PC	Storage tanks, boat hulls
PF	phenolic resins	$-C_6H_4-CH_2-O-CH_2-$ (ring with CH_2-)	PC	Electrical insulations, knife handles
UF	urea resins	$-NH-CO-N-CH_2-\overset{\mid}{N}-CO-NH-$	PC	Foams

that unnatural? They are certainly not "more" synthetic than iron, brass, or glass (see Chap. 1); some plastics even occur in nature. Storax (Chap. 4.1) and dragon's blood (from e.g. the Malayan rattan palm) are polystyrenes. A bug uses polymethyl methacrylate to immobilize its enemies (Chap. 5.1). There is even a mineral (elaterite) which is a highly branched polyethylene. What is so unnatural about plastics?

Plastics are not "plastics" but a big family of very different materials. Like metals, they come with many widely different properties. Like metals, they can also be subdivided into various groups, depending on manufacture or application. On distinguishes, for example, commodity plastics from specialty plastics and thermoplastics from thermosets.

Thermoplastics are plastics which become soft and shapable on heating and hard and solid on cooling. This transformation can be repeated many times without change of properties, similar to butter (not a polymer!) which softens at room temperature and hardens in the refrigerator. Thermoplastic polymers may be unbranched or branched; they may be crystalline like polyethylene, polypropylene and polyoxymethylene or amorphous like polystyrene and polymethyl methacrylate. A very high proportion of the world thermoplastics production (86%) comprises just four types of polymers: polyethylene (37,3%), polyvinyl chloride (27.1%), polystyrene (15.7%) and polypropylene (13.6%). Many other polymers share the remaining 6.3%.

Thermoplastics can be synthesized by the polymerization of carbon/carbon double bonds of monomers of the general type $CH_2=CHR$ (with $R=H$ in ethylene, $R=CH_3$ in propylene, etc.) or similar compounds

$$n\,CH_2=CHR \rightarrow \wedge\wedge(CH_2-CHR)_n\wedge\wedge \tag{11-1}$$

by polymerization of carbon/oxygen double bonds as in the polymerization of formaldehyde to polyoxymethylene

$$n\,CH_2=O \rightarrow \wedge\wedge(CH_2-O)_n\wedge\wedge \tag{11-2}$$

or by ring opening polymerization as in the polymerization of trioxane to polyoxymethylene

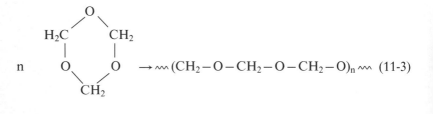

$$\tag{11-3}$$

or by polycondensation like the condensation polymerization of ethylene glycol with terephthalic acid (see Eq. (6-3)).

Thermosets, on the other hand, harden on heating and become never soft again, in the same manner that one cannot convert a cooked egg into a raw one. The hardening is caused by a crosslinking reaction in both cases, chemical crosslinking in thermosets, physical network formation in eggs (see also Chap. 8.5). Since the polymer can no longer be shaped after hardening, shaping and crosslinking must be performed simultaneously. The starting materials are usually low molar mass, multifunctional compounds, so-called resins. Polycondensation reactions are used most often, as for example for phenol/formaldehyde resins (Bakelite), for amino resins, or for polyurethanes. In a few cases, a linear polymer is formed first, which is then hardened by crosslinking polymerization. Examples are the unsaturated polyesters which are hardened by radical copolymerization with styrene

130

$$(11\text{-}4)$$

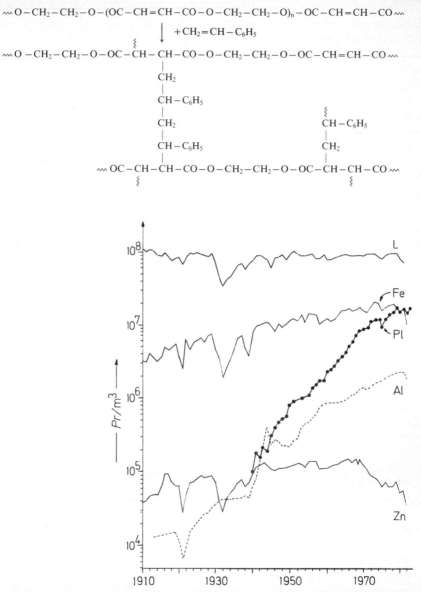

$$\small \sim\!\!\sim O - CH_2 - CH_2 - O - (OC - CH = CH - CO - O - CH_2 - CH_2 - O)_n - OC - CH = CH - CO \sim\!\!\sim$$

$$\downarrow \ + CH_2 = CH - C_6H_5$$

$$\small \sim\!\!\sim O - CH_2 - CH_2 - O - OC - CH - CH - CO - O - CH_2 - CH_2 - O - OC - CH = CH - CO \sim\!\!\sim$$

$$CH_2$$
$$|$$
$$CH - C_6H_5$$
$$|$$
$$CH_2 \qquad\qquad\qquad CH - C_6H_5$$
$$|$$
$$CH - C_6H_5 \qquad\qquad CH_2$$
$$|$$
$$\small \sim\!\!\sim OC - CH - CH - CO - O - CH_2 - CH_2 - O - OC - CH - CH - CO \sim\!\!\sim$$

Fig. 39. Annual production of various materials in the United States during the last 70 years. The production is given in cubic meters per year and plotted on a logarithmic scale. L = lumber, Fe = raw steel, PL = plastics, Al = primary aluminum (without recycled materials), Zn = zinc

131

Thermosets make up 20% of the plastics production. Similar to thermoplastics, four big groups of thermosets have captured 90% of the market: amino resins (39%), phenolic resins (23%), polyurethanes (15%), and unsaturated polyesters (13%).

Plastics consumption is quite different for various countries. Western Germany and Finland consume the most per capita, about 95 kg per annum. The corresponding numbers are 65 kg/a for the USA, 49 kg/a for Japan, 1.2 kg/a for African countries, and 0.4 kg/a for India. The world production of plastics has now reached 60 million tons, in part because of new plants in Eastern block countries, in rapidly developing countries like Brazil and Mexico, and in oil-rich countries like Saudi-Arabia. The production in highly industrialized countries has not stagnated either (Fig. 39); the United States still supplies about one quarter of the world's plastics production and West-Germany and Japan about one eighth each. The growth of the plastics production in industrialized countries is the more remarkable since the production of many other materials, like zinc, iron, and lumber, has declined (Fig. 39). Why are plastics so successful? The answer lies in their remarkable properties.

11.2 Stronger than Steel

People have their own ideas about plastics. Some think about soft and flexible films like garbage bags from polyethylene or shopping bags from plasticized PVC. Some remember the brittle and breakable plastics toys from standard polystyrene which gave the plastics industry a black eye several years ago. Some use the durable and beautifully shaped housewares from impact resistant polystyrene without thinking to much about their plastic nature.

These few examples indicate already the broad property spectrum of plastics. There are indeed plastics which are brittle and impact resistant, clear and hazy, flexible and rigid, burning and flame-resistant, electrically insulating and electrically conconducting, and many more. All have in common low densities ("specific gravities") of 0.9 to 1.5 g/mL, much lower than the densities of stainless steel (about 8 g/mL), aluminum (about 2.7 g/mL), cement (3.4 g/mL), or glass (2.6 g/mL). The low density saves considerable energy if plastics are used as materials for transportation (cars, trains, planes). Materials are sold per weight but are typically used per volume. This is the reason why the annual production of several materials was not given in mass units but as volumes in Fig. 39. The lower the density, the less material is needed for the same volume. If plastics are used in vehicles, lower dead weights result and the energy required for the movement of the vehicle is reduced. Cars contain already over 1000 plastics parts weighing more than 100 kg which replace at equal volume about 800 kg steel or 270 kg aluminum, a considerable saving of weight at car weights of 1000 to 2000 kg.

All fine and dandy. One can certainly manufacture dashboards, bumpers, and steering wheels from plastics. How about the essential parts, body and engine? The Ford Motor Company built already an experimental car engine from plastics but is is still undecided whether this plastics engine or the competing ceramics engine will win the race against engines from steel, aluminum or magnesium. And the body? Isn't the title of this subchapter too presumptuous? Plastics stronger than steel?

You read right. Plastics can indeed be stronger than steel and also stiffer. A measure for the strength of a material is the tensile strength (see Chap. 9.1). The scientific measure of the stiffness of a material is the modulus of elasticity which is given experimentally by the initial slope of the stress/strain curves (Chap. 9.1). Moduli of elasticity can also be calculated from molecular data and we will now estimate the theoretical modulus of elasticity of polyethylene from the known modulus of the diamond.

Diamonds consist of pure carbon. The carbon atoms are arranged regularly in a crystal lattice in which each carbon atom is connected tetrahedrally (Fig. 3) by covalent bonds with four other carbon atoms. The perspective representation of Fig. 40 does not allow to recognize that all four adjacent carbon atoms are at the same distance from a central carbon atom. One does see, however, that the carbon atoms are arranged in zigzag chains, just like the carbon atoms in chains of crystallized polyethylene (Fig. 20). The four adjacent "chains" of an arbitrary carbon chain in diamond thus correspond to the four hydrogen substituents of an ethylene unit $- CH_2 - CH_2 -$ in polyethylene. We know from spectroscopic experiments that the interactions between carbons and hydrogens practically do not contribute to the properties of the polyethylene chains. We can just concentrate on the carbon chains.

The modulus of elasticity of diamond in chain direction is 1160 GPa (for crystallographers: in [110] direction). Each carbon chain possesses a cross-sectional area of 0.0488 nm^2. Each polyethylene chain, on the other hand, exhibits a cross-sectional area of 0.182 nm^2. The modulus of elasticity of polyethylene in chain direction should thus be at least $E = 1160 \cdot 0.0488/0.182$ GPa $= 310$ GPa. Exact theoretical calculations indicate a theoretical modulus of 340 GPa in chain direction and one of 3 GPa perpendicular to the chain. Conventionally processed polyethylenes have however moduli of less than 1 GPa.

The difference between the theoretical and the experimental moduli is caused by the different arrangement of the polyethylene molecules. Poly-

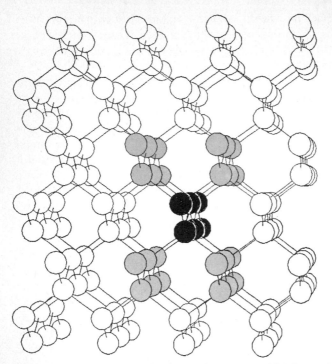

Fig. 40. Section of a diamond lattice. Carbon atoms (white) form "chains"; each individual chain (e.g., in black) is surrounded by four adjacent chains (gray)

ethylene adopts more or less a coil conformation in the melt. On cooling, chain segments cannot reach their ideal lattice positions fast enough and some remain in the coil conformation. The chain axes of these partially crystallized polyethylene molecules are distributed in all directions. In the theoretical model, all chains must be totally crystallized and furthermore aligned in only one direction.

Several research groups have succeeded to prepare polyethylenes with high moduli by various methods. Moduli of 260 GPa have been already reached in laboratory experiments. Three companies have constructed pilot plants for polyethylene fibers with moduli of about 100 GPa.

The stiffness of a polyethylene with a modulus of elasticity of 260 GPa is already higher than that of steel of 210 GPa. But since the volume and not the mass is important for the application of materials, these moduli should be divided by the corresponding densities. Steel has a much higher density than polyethylene and so the "specific" modulus of steel is thus about ten times smaller than that of the best polyethylene in chain direc-

Table 26. Moduli of elasticity and tensile strengths at break of various materials

Material	Density in g/mL	Modulus in GPa	Tensile strength in GPa	"Specific" modulus in GPa·mL/g	"Specific" strength in GPa·mL/g
Tungsten	19.3	410		23	
Steel	8.0	210	4.3	26	0.54
Aluminum	2.8	70	2.1	25	0.75
Glass	2.6	70	3.6	27	1.38
Diamond	3.51	1160		330	
Graphite	1.66	730	20	440	12.0
Polyethylene[a]					
chain direction	1.00	260	20	260	20
perpendicular to chain	1.00	3	0.2	3	0.2
Poly(*p*-phenylene terephthal-amide)[a]	1.44	150	2.8	104	1.9

[a] Best experimental data; conventional polyethylenes have far lower values.

tion. Even the very stiff tungsten metal has a specific modulus of only 23 GPa·mL/g, compared to polyethylene's 260 GPa·mL/g.

Similar considerations apply to the strengths of materials. The tensile strength at break of ultra-oriented polyethylene is about 5 times that of steel and its specific tensile strength is even 40 times higher than that of steel (Table 26). Polyethylene can indeed be stronger than steel.

How about the other polymers? Do they have similar properties as polyethylene? Yes and no, since the properties depend on chemical and physical structures. We saw already for the estimation of the modulus of elasticity of polyethylene from that of diamond that the cross-sectional area of chains is very important. Moduli and strengths are area-related forces. The greater cross-sectional area means that fewer chains are available per area to distribute forces, with a resulting lowering of the moduli and strengths of otherwise similar structures. "Similarity" refers here to the chain conformation. Zig-zag chains are rather rigid "rods". Their extension in chain direction requires increases in valence angles and bond lengths which require lots of energy. Helices, on the other hand, are spirals. Their extension requires only the transformation of gauche conformations into trans conformations which requires much less energy. Polymers in helix conformation are thus less stiff than polymers in zig-zag chains.

Examples of zig-zag chains with high theoretical moduli are polyethylene (340 GPa), polyamide 6.6 (200 GPa), polyethylene terephthalate (150 GPa) and cellulose (as flax; 130 GPa). Examples of theoretical moduli for helical chains are isotactic polypropylene (50 GPa) and polyethylene oxide (9 GPa). The moduli of polymers in their conventional applications are much lower, for example, 1 GPa for polyethylene, 5 GPa for polyamide 6.6, and 3 GPa for isotactic polypropylene.

11.3 Unity Creates Strength

The high stiffnesses and strengths of ultra-oriented polymers are caused by the forced parallelization of chains of flexible macromolecules (Fig. 41). All synthetic fibers reach their high strengths only if they are further "drawn" after spinning. The ultra-oriented polymers result correspondingly from ultradrawing.

High strengths are much more easily obtained from rigid rod-like polymer molecules, for example, with the poly(p-phenylene terephthalamide) of Fig. 19. The long axes of these molecules are arranged at random in

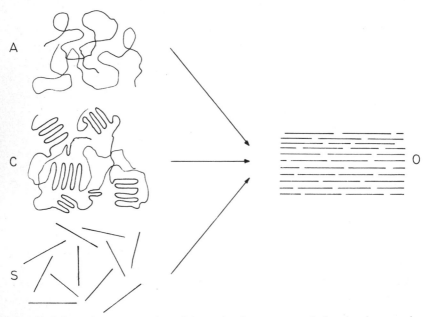

Fig. 41. Schematic representation of the molecular processes during drawing or spinning of flexible polymers in the amorphous state (A) or the partially crystalline state (C) or the processing of rigid rod-like polymer (S) to oriented polymers (O). Polymer chains are given as lines without consideration of valence angles

their dilute solutions at rest (Fig. 41 S). On spinning of these solutions through nozzles, molecules align themselves parallel to the flow direction, similar to a raft of logs in a river. The molecules remain in that arrangement after rapid removal of the solvent. The resulting Kevlar fibers are stiff and strong and are used, for example, for bullet-proof body armor.

If rigidity and strength of materials can be improved by the parallelization of molecules, then less strong materials should be made stronger by the addition of fibers. Such composite materials are already used in great amounts. Best known are probably the glass-fiber reinforced unsaturated polyesters, popular for boat hulls. Unsaturated polyesters themselves are only low strength/low modulus materials. Their mechanical properties increase considerably after blending in short glass fibers or if they are embedded in glass fiber mats.

Polymer composites are especially popular for high technology applications. Glass fiber reinforced epoxy resins achieve much higher tensile strengths at break than iron, aluminum or even titanium. Over 15 tons of weight were saved by using polymer composites in the new Boeing 757 plane. This saved weight corresponds to additional 15 tons of freight or 150 additional baggage carrying passengers or corresponding savings in fuel.

No reason to get excited, by the way. Nature has known the principle for millions of years and our ancestors used it in many ways. Wool is a composite of bundles of parallel fibers in an amorphous matrix (Fig. 33). Wood is an especially sophisticated composite: an amorphous matrix of cross-linked lignin molecules is reinforced in growth direction by cellulose fibers. Wood is very rigid in the growth direction, but can be easily compressed perpendicular to it. The low density of ca. 0.5 g/mL of many woods is achieved by a kind of foam structure; lumber floats on water (density of ca. 1 g/mL) until the included air is replaced by water and the wood sinks to the bottom of the waters. The cellulose of wood is furthermore plasticized by water (see Table 24); dried wood is fairly brittle and cracks easily. Some woods, for example teak, are plasticized by oils and not by water.

Our ancestors knew the principle of fiber reinforcement, too. 5000 years ago, boats were built in Mesopotamia and Egypt from reed bundles which were embedded in bitumen, a forerunner of the boats from glass fiber reinforced polyester resins. Mummies were prepared in Egypt by the wrapping of corpses in linen fabrics which were then impregnated with natural resins. We now use a similar technology for the manufacture of jet fans and for storage tanks from polymer impregnated fiber mats. Bricks from loam were reinforced by reed mats in Mesopotamia 4000 years ago, a forerunner of steel-reinforced concrete.

Steel-reinforced concrete is a concrete which is reinforced by iron rods or wires; concrete itself is a composite of cement, gravel, sand and water and has a very low tensile strength without steel reinforcement. Cement is manufactured from limestone, clay (an aluminum silicate), iron oxides, and magnesium carbonates.

Cement and water "bind" under formation of cement glue, cement gel, and finally cement stone: a cross-linking polycondensation to silicates. During the binding, fiberlike silicate and aluminate crystals grow from the cement particles. They form a fleece which embedds gravel and sand. The binding proceeds for weeks; the final strength is only reached after three months.

Concrete does not stop to react even then. The non-reacted lime of the concrete reacts in air with carbon dioxide to limestone and with sulfur trioxide to gypsum. The concrete loses its basicity and becomes neutral. Limestone and gypsum expand during their formation and create pores. Water penetrates the pores. It expands on freezing and cracks the concrete. Steel is then no longer protected by the concrete. Water and oxygen of the air can now attack the steel and form rust. Since rust is less dense than concrete, new cracks are created. The process is vastly accelerated by dynamic loading; concrete creeps under load which promotes the formation of grooves which ultimately cause aquaplaning. A truck of 25 tons weight causes the same damage to bridges and roads as 6500 cars. Shouldn't the tax be 6500 times as high for such vehicles?

Certain organic polymers may be used to prevent cracks or repair concrete structures. These polymers seal the pores so that water can no longer penetrate the concrete. Silicone resins are especially suited for concrete sculptures and monuments.[1] Also useful are "polymer cements", which are blends of inorganic cement with organic polymers.

---

[1] For a more detailed explanation and further examples, see: E. Rochow "Silicon and Silicones".

12 Everything Flows

The mountains gushed before the Lord.

Judges 5,5 (The Song of Deborah)

12.1 Deborah and the Tennis Rackets

Everything flows according to Heraclitus (544–483 B.C.), why not the mountains, too? Before the Lord, not before us, since they flow only in geological time ranges. We consider rocks and stones as the prototypes of solid materials ("solid as a rock"), just as well as iron, glass, rock salt, etc. These materials do flow, however. Rock salt formations in salt mines deform slowly under the pressure of the rocks above them. Old window panes in churches are thicker at the bottom than at the top since glass will flow under its own weight. Engine hoods deform slowly with time if the steel sheets used are too thin.

All these materials are considered "solids" since they do not deform markedly under the "usual" loads in "usual" time spans. A steel sphere bouncing on a steel plate forces the iron atoms to move a little from their resting positions in the crystal lattice. The resulting deformation causes a stress since the atoms like to return to their energetically favored resting positions. This stress causes the spheres to bounce back (rebound). The iron atoms again occupy their original positions if the deformation was smaller than about 1%. Such bodies are called "ideal elastic" or "energy elastic".

Large and long-lasting deformations cause energy elastic materials to deform irreversibly. Car panels and bodies deformed in accidents cannot be brought back to their exact old shape even by the most skilled mechanic. Even "solids" flow if a sufficient high load acts long enough.

Liquids seem to behave totally different. They deform "immediately" even under small and short-time loads. Liquids always assume the shape of the surrounding container. They never maintain any shape without external encouragement. Liquids always flow under their own weight and are viscous materials.

Viscous and energy-elastic behaviors are only two limiting types of behavior of matter according to rheology, the science of flow phenomena (Greek: rheos = current, stream). Every energy-elastic body is also somewhat viscous; steel would otherwise not deform under big and long-lasting loads. Each viscous body, on the other hand, is also somewhat energy-elastic.

The two extremes of rheological behavior can be described by simple mechanical models. An energy-elastic body behaves like a spring according to R. Hooke (1635–1703) (Fig. 42). A load compresses the spring "spontaneously". The spring stays in the compressed state until the load is removed. It then returns "immediately" to its original position. "Spontaneous" and "immediately" mean only that deformation and relaxation times are very, very small compared to the observation times. A plot of the deformation against the time thus exhibits the typical square-box diagram of such a Hookean body.

The deformation of a viscous liquid can be described by a dashpot. The plunger tries to move through the liquid under load but the liquid resists the movement and the plunger moves slowly with time. A certain deformation is reached after a while. Removal of the load causes the plunger to stay in this position.

Springs and dash-pots can be combined in various ways: in series according to Maxwell and in parallel according to Voigt and Kelvin (Fig. 42). A "sudden" load causes a "sudden" initial deformation of a Maxwell body followed by a slow movement of the dashpot through the

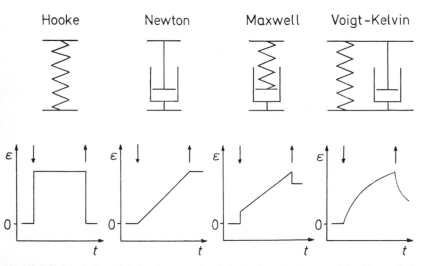

Fig. 42. Mechanical models for the stress/strain behavior of polymers. A load is applied at ↓ and the body is deformed. The load is removed at ↑. The deformation ε is a compression under pressure and an elongation under tension. Only the Hookean spring allows instantaneous deformations

liquid. Removal of the load lets the spring retract but the dash-pot stays put, and a permanent deformation results.

The deformation is opposed by a stress: a compressed spring is under stress but so is an expanded spring. The stress is directly proportional to the deformation in the Hookean body; removal of the stress lets the deformation disappear immediately and vice versa.

The Maxwell body behaves differently: the stress at a fixed deformation decreases with time. This process is called a "relaxation" in mechanics. How different from human behavior where a relaxation is supposed to remove both the stress and the deformation. But does it really? Ask your psychiatrist.

The relaxation time is the time after which the stress is reduced to $1/2.718 \ldots$ of the original value (1 divided by Euler's number "e"). The ratio of relaxation time to the observation time is called the Deborah number after the prophetess of the Old Testament. The Deborah number is quite appropriately zero for liquids, unity for polymers at the glass transition temperature, and infinity for ideal elastic bodies.

These dynamic properties of polymers affect many applications, for example, the behavior of strings of a tennis racket. The strings deform on impact of the tennis ball and then return to their old position. The magni-

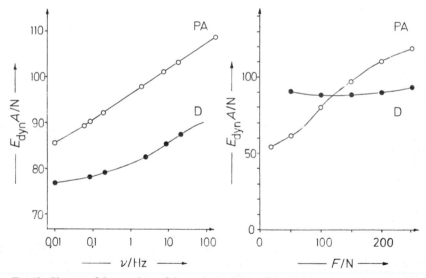

Fig. 43. Change of the product of dynamic modulus of elasticity and area (measured in newtons) with frequency ν (measured in Hertz) or pre-load F (measured in newtons) for tennis strings from polyamide (PA) or catgut (D) (after data of J. Koppelmann)

tude of the deformation depends on the tension put on the strings (pre-stress, pre-load), the dynamic modulus of elasticity of the strings, and the residence time of the ball on the racket.

The dynamic modulus of elasticity is an area-related force. Since strings possess different diameters, dynamic moduli must be multiplied by areas for a comparison of forces. The strings are subjected to these dynamic forces, for example, in a more-or-less sinusoidal manner. A residence time t then corresponds to a frequency $v = 1/(2\ t)$. The factor $1/2$ arises from assuming the deformation to be half of a sine wave. A residence time of $t = 0.005$ s thus corresponds to a frequency of $v = 1/(2 \cdot 0.005)$ s^{-1} $= 100$ Hz.

Strings from catgut and polyamide (both polymers) behave quite differently as every tennis player knows. The dynamic force increases with increasing frequency for both types of strings (Fig. 43, left). All strings thus become stiffer with increasing frequency or decreasing residence time of the ball. Tennis players say that the rackets become "harder" with fast balls or "have more pace". Polyamide strings are always harder than catgut strings; they allow a faster play.

Catgut strings maintain their "hardness" over a wide range of preloads whereas the "hardness" of polyamide strings increases with increasing pre-load (Fig. 42, right). Rackets with catgut strings can thus be stringed with slightly different pre-loads at various parts. No such variation is permitted for polyamide strings, however, since different pre-loads at different parts of the rackets would cause different hardnesses and the flight trajectory of the ball would become much more incalculable. Polyamide strings also relax a little with time. Their pre-load and thus also their hardness decrease slowly under constant deformation. Polyamide strings therefore have to be adjusted much more often than catgut strings.

12.2 Entangled Knee-Joints

As mentioned, a "relaxation" is the decrease of stress with time in the language of mechanics. An increase of deformation with time at constant stress is called "creep", and characterized by a "retardation time". "Creep" and "after-flow" of a material can be represented by the Voigt-Kelvin model (Fig. 42).

Spring and dash-pot are parallel to each other in this model. The spring wants to deform immediately under load but it cannot for it is restrained by the dash-pot. The deformation thus increases only slowly with time, not instantaneously. The removal of the load likewise leads to a slow recovery to a final value.

Creep is observed with polymers even at room temperature. This phenomenon seemed very mysterious at the time of its discovery. The plastics known at the time assuredly looked "solid"; one could manufacture beakers and containers which showed no sign of softness. Yet, O-rings and washers made from these polymers did not hold tight under pressure since the materials flowed away even at room temperature. They showed "cold flow". But why?

Mechanics was not able to answer this question since it can relate various physical properties to each other but does not attempt a molecular interpretation. The "why" of mechanics is not the different "why" of the molecular sciences. Look at the different answers when you ask why somebody broke his leg. The layman will say because he was so unfortunate to fall. The engineer because he slipped on a banana peel. The lawyer because somebody threw away the banana peel in spite of it being forbidden to litter. The physician because the bones were not strong enough...

Why do thermoplastics creep? Because they consist of chain molecules. Because their chains are not crosslinked. Because only small intermolecular forces are present. Because they contain amorphous regions. Because their free volume allows a certain flexibility of the chains. Because ... Confused?

Chain atoms can move fairly rapidly above the glass transition temperature from one conformation into the other (Chapt.7.1) but very slowly below this temperature. A load forces certain chain atoms to adopt other conformations, just as grass blades bow under the weight of a human. Neighboring chain molecules are of no help either since only weak forces exist between them, and which cannot help maintain the original spatial positions. The chains slip along each other and flow under the load. Not as fast as spilled water but faster than flowing mountains. The flow is nonetheless so "fast" that thermoplastics cannot be used to build bridges.

Is there nothing which can stop this flow? Oh yes, quite a number of things. Creep and cold flow depend on the ratio of retardation time to the time under stress. A high ratio lets the material behave like a rigid solid. It will not creep and hence it maintains its shape. Containers from polystyrene are good examples but even these will creep if kept under load long enough.

The ratio retardation time/stress time can sometimes extend to ranges which can be observed under conventional loads. Garbage bags from polyethylene are examples. Such bags are expected to stretch around their sometimes edged contents but they should not flow away. Polyethylene performs this miracle by a "physical crosslinking". Some chain segments of polyethylene are crystallized, that is, firmly anchored in a crystal lattice. These segments will no longer slip away under modest loads. Since one chain may reside in several crystalline regions (Fig. 40 C), segments between such regions will not move easily. The ability to stretch is present, but only to a limited extent.

Partially crystallized polymers like polyethylene can thus be considered physical thermosets. Conventional polyethylenes are, however, not near as temperature stable as true thermosets. Polyethylene itself exhibits a heat distortion temperature of only 50 °C under a load of 1.89 MPa, despite melting temperatures of 130 °C (conventional PE) to 145 °C (ultrastretched PE). The heat distortion temperature characterizes the temperature at which a rod bends to a certain extent under a load of 1.89 MPa and a heating rate of 2 K/min, and is thus a measure of the tendency to creep.

Conventional polyethylene tends to creep under higher loads and can therefore not to be used for artificial knee and hip joints. In these knee joints, a spherical ball from a special steel moves in a socket of ultrahigh molar mass polyethylene (3 000 000 g/mol versus 100 000 g/mol for conventional polyethylene). High molar mass polymers have more repeating units per chain than low molar mass compounds and since each repeating unit can exert weak attraction forces to other units, a high molar mass polymer exhibits many of these contacts per chain. It is thus very difficult to severe all contacts simultaneously as would be required for a flow or creeping process.

A second effect is even more important. Very long chains entangle (Fig. 44). Short-range forces can not remove the loops, on the contrary, loops now act as physical crosslinks. We all made that experience with coils of strings: the more we pull, the more the strings become entangled. However, very slow and very patient pulling will disentangle the strings.

How does that happen since one high molar mass polymer chain is surrounded by so many others and takes part in a multitude of loops? Probably the same way that a snake moves when it glides through a brush pile. The loops form a kind of tube through which the chains "reptate" (Fig. 45).

The entanglements raise the heat distortion temperature of ultrahigh molar mass polyethylene to 115 °C from 50 °C for conventional polyethyl-

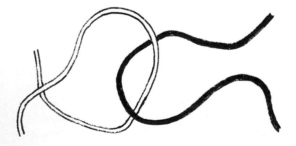

Fig. 44. Entanglement of two chain molecules

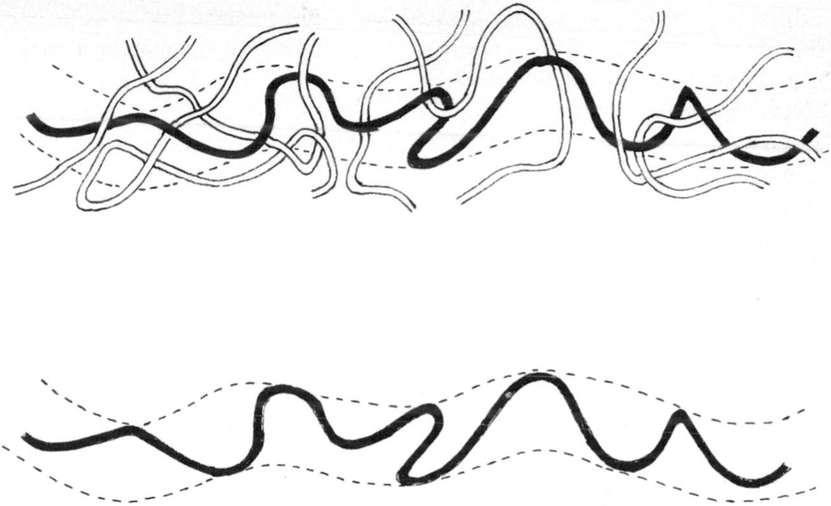

Fig. 45. Entanglement of a chain molecule (black) by segments of other chain molecules (white). The chain "reptates" through a kind of tube (dotted lines)

ene. High molar mass polyethylene can be used for highly stressed artificial joints. But, for the very same reason (resistance to flow), ultrahigh molar mass polyethylenes can no longer be processed by injection molding or extrusion; they have to be shaped by metallurgical techniques, e.g., forgery.

12.3 Rubbers, Gums, and Elastomers

Chemical crosslinking via covalent bonds is the easiest way to prevent the creep and flow of polymer chains. All chains become connected with each other and the whole specimen becomes one giant macromolecule (Chap. 5.4). Crosslinking reactions were mentioned several times in this booklet: for thermosets (Chap. 11.1), leather (Chap. 9.3), printing plates and resists (Chap. 6.3), and permanent waves (Chap. 10.2).

Chemical crosslinking of polymers is an old art which was performed empirically without knowledge of the molecular processes. Various trades used different names for what is now known to be the same process, albeit with different chemical reactions. The crosslinking transformation of skins and hides into leather was called "tanning", the crosslinking polymerization of surface coatings based on linseed oil was "drying", the radiation crosslinking of other coatings was "curing", the crosslinking of natural rubber to elastomers was "vulcanization", the crosslinking of resins to thermosets was "hardening" and the crosslinking of hair in the permanent-wave process was "fixation".

The molecular process common to all these phenomena was not recognized for many years. After all, why should the hardening of a brittle resin to a rigid thermoset be related to the transformation of a sticky rubber to an elastic material?

Common to all these processes is the formation of crosslinks between the chains. The chemical nature of these crosslinks is of course influenced by the crosslinking process, that is, the various chemical reactions involved, but it does not greatly affect the properties of the resulting crosslinked product. These properties are much more influenced by the crosslinking density (the number of crosslinks per unit volume) and the lengths of the crosslinks themselves.

Highly crosslinked networks have very short connecting chains between the primary chains (Fig. 46). All chain segments are restricted in their mobility because they are hooked to their neighbors and the crosslinked polymer is rigid and stiff both below and above the glass transition temperature. Lightly crosslinked networks are wide-meshed and the chain segments between the crosslinking points move relatively freely. How free, depends on the glass transition temperature. If the use temperature is below the glass transition temperature, then the crosslinked polymer behaves as a thermoset. However, if it is above the glass transition temperature, then the polymer behaves like an elastomer.

The crosslinking density directly affects the end-use properties. Surgical gloves should be very "soft" and flexible, and so chain segments must therefore be able to move relatively freely under stress but they should, of course, not flow away. Surgical gloves therefore consist of lightly crosslinked polymers above their glass transition temperatures; there are about 100–150 isoprene repeating units between the crosslinking points in surgical gloves from vulcanized natural rubber. Household gloves from rubber are less soft and have 50–80 repeating units between crosslinks. Tire treads must be highly elastic but not too deformable; the crosslinking density is reduced to 10–20 repeating units. Hard rubber is no longer an elastomer since there are only 5 or so isoprene units between crosslinking points.

Natural rubber no longer dominates the rubber industry; its share has sunk to ⅓ of the total world rubber consumption. The world production of synthetic rubbers is now about 6.4 million tons, of a nameplate capacity of about 9.2 million tons. Eastern block countries each year perform a technical miracle: they overfullfil the plan which equals the name plate capacity. In Western countries, production (5.4 million tons) is significantly lower than theoretical capacity (8.2 million tons). The reason is the changing

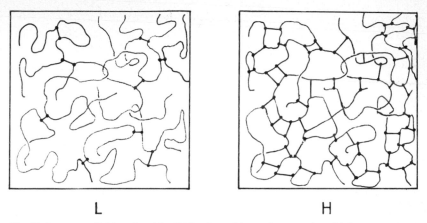

Fig. 46. Low crosslinking densities (L) lead to wide-mesh networks with large segments between crosslinking points, high crosslinking densities (H) to narrow-mesh networks with short segments

consumer behavior. The actual capacity of a chemical plant is always lower than the (theoretical) name plate capacity since there must be dead times (cleanings, bad charges, and so on) even if the plant is not on strike; the actual capacity is usually 90% of the theoretical. The much lower actual production of 66% of the theoretical capacity for synthetic rubber must therefore have other reasons.

The United States produced in 1974 almost 2.6 million tones of synthetic rubber. The lion's share went to styrene-butadiene rubber (56.5%), with the rest made up by butadiene rubber (12.1%), chloroprene rubber (6.4%), butyl rubber (6.4%), ethylene-propylene rubber (4.9%), synthetic isoprene rubber (3.6%), nitrile rubber (3.6%) and others (5.7%). Styrene-butadiene rubbers are mainly used for automobile tires, often in a blend with natural rubber.

The first oil crisis in 1974 reduced car production and cars were also driven less. Styrene-butadiene rubber production promptly declined in 1975 (Fig. 47). It recovered in 1976-1979 but now buyers showed their love for smaller cars with better gas mileage and smaller tires and especially to imported cars which brought their tires with them. Less and less styrene-butadiene rubber has been produced in America ever since. Other synthetic rubbers were hit neither by the recession nor by the changed consumer behavior.

Rubbers are rarely used "pure", i.e., without addition of other chemicals. Surgical gloves are an example. In general, the optimal rubber properties are only achieved by formulation with other chemical compounds.

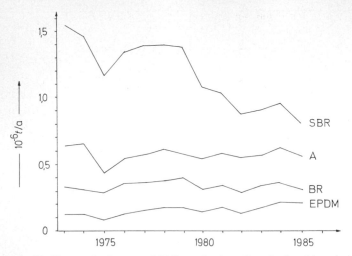

Fig. 47. Changes in the annual U.S. production of synthetic rubbers between 1973 and 1985 (in million tons per year). SBR = Styrene-butadiene rubber, BR = butadiene rubber, EPDM = ethylene-propylene rubber, A = other synthetic rubbers

The result is not too dissimilar to the preparation of grog: rum makes it strong, added water less strong, lemon makes it sour and sugar sweet.

Formulated rubbers are fairly complicated mixtures. Rubber is first masticated (Chap. 2.1) to lower molar masses. The resulting viscosity allows a simple, homogeneous mixing in of additives which is promoted by the continuous creation of new surfaces.

Rubber provides elastomers with the desired visco-elastic properties. The vulcanization process with, for example, sulfur as crosslinking agent transforms the "viscous" rubber into the "elastic" elastomer. The glass transition temperature is sufficiently increased by the crosslinking to warrant a plasticization of the elastomer by mineral oil which causes the glass transition temperature to decrease. Did you know that some elastomers contain up to 40% mineral oil? Addition of mineral oil reduces the tensile strength at break, however. Carbon black, a so-called active filler, is therefore added since it improves the mechanical properties.

The properties of elastomers vary with rubber type, additives, and vulcanization. Elastomers as a group are however quite distinct from thermosets (Table 27). Segments between crosslinking points remain flexible in elastomers because of their relatively long chain lengths and because of the low glass transition temperatures. On tension, chain atoms can thus adopt new, more extended conformations; the population of trans conformations increases. The elongation at break can be very high. Since the

Table 27. Properties of some important, unfilled elastomers and thermosets

Polymer		Glass temperature in °C	Modulus of elasticity in MPa	Resilience in %	Elongation at break in %
Symbol	Name				
Elastomers					
NR	Natural rubber	−73	5[a]	40	780
BR	Butadiene rubber	−95	7[a]	65	510
SBR	Styrene-butadiene rubber	−52	9[a]	40	650
EPDM	Ethylene-propylene rubber	−55	9[a]	45	320
IIR	Butyl rubber	−63		12	800
Thermosets					
PF	Phenolic resin	75	6300	0	1
UP	Unsaturated polyester	>60	3400	0	5
EP	Epoxy resin	>45	2500	0	5
PUR	Polyurethane	90	28	1	5

[a] At 300% elongation.

stretched rubber contains more trans conformations than corresponds to equilibrium, removal of the stress lets the rubber adopt again its equilibrium population of conformations. The rubber thus jumps back to its initial state without being permanently deformed.

Compression is the opposite process. Again, equilibrium populations of conformations are shifted, this time to more gauche conformations. Unloading restores the equilibrium populations and the rubber bounces back. The resilience of rubbers is often around 40% (Table 27), i.e., if a rubber ball is dropped from a certain height, it bounces back to about 40% of its original height. Butyl rubber has exceptionally low resilience and is not used for tire treads because of relatively poor wear. Since it has a very low permeability to gases as compared to natural rubber and styrene-butadiene rubber, it is used for inner tubes.

12.4 Physical Vulcanizations

Rubber tires are difficult and expensive to manufacture. Mastication of baled rubber requires heavy equipment and the working in of additives is time consuming. The formulated rubber is then shaped into stripes and patches which are layered on drums. The raw tire receives its tread design and actual shape in a pressure autoclave where it is heated to the vulcanization temperature.

This (almost) stone age technology is capital, labor, and energy intensive; the chemical crosslinking process consumes time and produces worthless scrap. Mixing and shaping of low viscosity masses without vulcanization would thus be a worthwile goal.

Liquid rubbers were the first to be tried out. Liquids are much more easy to mix with other liquids (plasticizers) and powders (fillers, sulfur, etc.) than tough bulk rubber. The goal was to process the viscous mixture by injection molding instead in a heavy pressure autoclave. The liquid rubbers still have to be vulcanized, however.

Rubbers must have very low molar masses of about 10000 g/mol in order to be "liquid". The lower the molar mass, the more chain ends are present per unit volume. Many of these chain ends will not be incorporated into the network during the vulcanization. They will dangle freely and will not add anything to the elastic properties of the network. Hence liquid rubbers were found not to produce good tires.

But why use liquids if one can use powders? Powders can be as easily mixed as liquids. The molar masses of powdery materials may be very high since it is the viscosity of the powder particles and not that of the particle molecules which affects the ease of mixing. Molar mass is thus not a limiting factor for the use of powdered rubber.

Rubbers have one other outstanding property, however: they show tack. Freshly cut pieces of unvulcanized rubber will stubbornly stick together after being pressed together for only a few seconds. The tack is helpful for traditional tire manufacturing but not at all for the preparation of powdered rubbers.

Why do rubbers stick? Rubber molecules are above their glass transition temperatures at room temperature (see Table 27). Chain atoms can thus move easily from one conformation into another, especially under pressure. When two rubber pieces are pressed together, the molecules diffuse into each other and the two rubber pieces become anchored by rubber molecules which possess segments in both pieces.

The diffusion of molecules is measured by the diffusion coefficient. The self-diffusion coefficients D of rubber molecules above the glass transition temperature are approximately 10^{-12} cm^2/s. Typical rubber molecules exhibit coil diameters of ca. $L = 40$ nm $= 4 \cdot 10^{-6}$ cm. A penetration depth of one coil diameter would be sufficient to anchor chains firmly in both rubber pieces. The necessary time t can be estimated via the Einstein equation

$$D = L^2/(2\ t) \tag{12-1}$$

as $t = L^2/(2\ D) = (4 \cdot 10^{-6}\,\text{cm})^2/(2 \cdot 10^{-12}\,\text{cm}^2 \cdot \text{s}^{-1}) = 8$ s. Even if we erred by a decade of time: the time to achieve good tack in rubbers is still in the range of seconds, not days.

Powdered rubbers can not be prepared by grinding since the rubber particles will stick together immediately. However, the problem was solved by the precipitation of dissolved or dispersed rubber in the presence of inert fillers. The fillers stick to the surface of the precipitated rubber particles and provide a non-sticky coat. Such powdered rubbers are more and more being used instead of baled rubber.

Liquid and powdered rubbers do not solve the second problem of the traditional elastomer technology: the expensive and time-consuming chemical vulcanization. To get around this problem, polymer scientists came up with something very clever: thermoplastic elastomers. These polymers can be processed like thermoplastics but exhibit elastomer properties after cooling to room temperature. They vulcanize physically, not chemically; all scrap can be reused since it is not chemically modified as in conventional (chemical) vulcanization.

The development of thermoplastic elastomers is an example how the chemical and physical understanding of materials works together to create a new technology:

Amorphous polymers adopt coil conformations both above and below the glass transition temperature. Suppose coils of A polymers with n A units which have the same unperturbed dimensions as coils of B polymers with m B units. A mixture of polymer A_n with polymer B_m will be incompatible in most cases; it tries to separate in two immiscible phases, one with all the A_n molecules, the other with all the B_m molecules. What happens if one A_n molecule is coupled to a B_m molecule? The A_n blocks of these A_n-B_m diblock molecules still want to cluster together and separate themselves from the phase with the B_m blocks. Such a complete separation of all the A_n and B_m blocks is however no longer possible since the blocks are firmly bound to each other. Each block tries to preserve its unper-

A lamella

Block
molecule

A_n A_n A_n A_n A_n A_n A_n A_n A_n A_n A_n A_n
| | | | | | | | | | | |
B_m B_m B_m B_m^- B_m B_m B_m B_m B_m B_m B_m B_m

B lamella

Block
molecule

B_m B_m B_m B_m B_m B_m B_m B_m B_m B_m B_m B_m
| | | | | | | | | | | |
A_n A_n A_n A_n A_n A_n A_n A_n A_n A_n A_n A_n

A lamella

Block
molecule

A_n A_n A_n A_n A_n A_n A_n A_n A_n A_n A_n A_n
| | | | | | | | | | | |
B_m B_m B_m B_m B_m B_m B_m B_m B_m B_m B_m B_m

B lamella

turbed dimensions. Result: A_n blocks arrange themselves in layers facing their chemically coupled B_m blocks on one side and A_n blocks of other A_n-B_m block molecules on the other (Fig. 48). The polymers exhibit a layer or lamella structure if the unperturbed dimensions of the B_m blocks equal those of the A_n blocks (p. 151 and Fig. 48 L).

Great gaps would however occur between the coils of the A_n blocks in a lamella structure, if the unperturbed dimensions of the B_m blocks are much greater than those of the A_n blocks (see Fig. 48 S, left). Nature abhors vacua and the block polymers are therefore given the choice of either maintaining the lamella structure and abandoning the unperturbed dimensions (by spreading the A_n blocks or by compressing the B_m blocks) or just the other way around: preserving the unperturbed dimensions and arranging the coils in a different kind of morphology. The first choice is energetically less favorable and the coils of the B_m blocks thus fill the space by arranging themselves around the coils of A_n blocks (Fig. 48 S, right). Such block polymers exhibit cylindrical or spherical domains of coil-like A_n blocks in a continuous matrix of coils of B_m blocks (Fig. 48 T), depending on the relative coil size of A_n and B_m.

The same morphology is obtained for triblock polymers A_p-B_m-A_p (with $p = n/2$) as for diblock polymer A_n-B_m. The two A_p blocks of one tri-block polymer molecule belong however to two different domains which is equivalent to the physical crosslinking of the B component. If now the glass transition temperature of the A_p blocks is above the use temperature and that of a continuous matrix of the B_m blocks below, then the triblock polymer behaves like a physically crosslinked rubber. The crosslinking domains "melt" above the glass transition temperature and the triblock polymer can be processed like a thermoplastic. On cooling below the glass transition temperature, they form "hard" domains in a "soft" matrix of the rubbery B_m blocks. A commercial example is the triblock polymer $(sty)_p$-$(bu)_m$-$(sty)_p$ with styrene end-blocks (glass transition temperature ca. 80 °C) and butadiene center blocks (glass transition temperature ca. -95 °C). This polymer is thus "crosslinked" by the "hard" domains of styrene blocks.

Diblock and triblock polymers are examples of "self-organizing" polymers: lamellae are obtained if the different blocks possess more-or-less equal unperturbed dimensions whereas one observes spheres of the minor component in a matrix of the major component if the unperturbed dimensions are very different. If the ratio of the unperturbed diemensions is somewhere between these two extremes, cyclindrical domains of the minor component in a continuous matrix are observed.

This beautiful technology has however two shortcomings. Triblock polymers as thermoplastic elastomers are not as easy to manufacture as

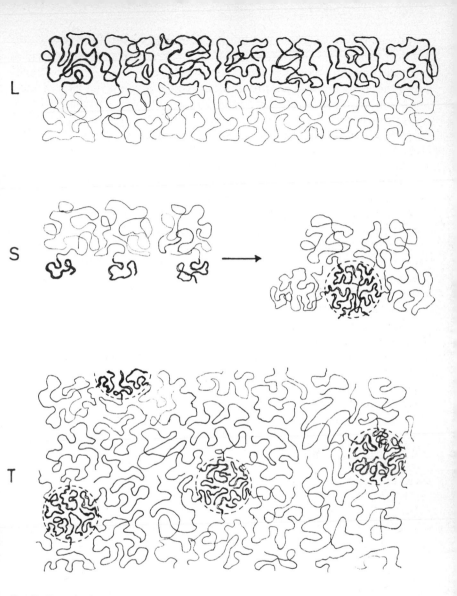

Fig. 48. Domain formation as a result of various relative coil dimensions of polymer blocks A (fat lines) and B (thin lines) of two-block polymers $A_n - B_m$. L = Lamellae from equal unperturbed dimensions of A_n and B_m. S = Spheres of A blocks in a continuous matrix of B blocks if the A_n blocks exhibit much smaller unperturbed dimensions than the B_m blocks. T = Thermoplastic elastomers with spherical (crosslinking) domains from A blocks of triblock polymers $A_p - B_m - A_p$

are conventional rubbers and thus are more expensive. A triblock synthesis can follow various strategies. One can first prepare an A block, attach a B block by the method of living polymerization (Chapt. 5.3) and finally cap the molecule with another A block. Unfortunately, not all molecules are willing to follow the good intentions of chemists and the method is thus restricted to a few monomer pairs. Another strategy synthesizes an A block and then attaches to this block a half B block. Two molecules $A_p - B_{m/2}$ are then coupled with the help of a small C molecule to the desired $A_p - B_{m/2} - C - B_{m/2} - A_p \approx A_p - B_m - A_p$.

Much more serious is another birth defect: the low glass transitions temperature of the polystyrene blocks of maximal 100 °C. These thermoplastic elastomers cannot be used for the manufacture of rubber tires since tires heat up to 90 °C on cruising speeds of 100–120 km/h (63–75 mph) and 120–130 °C at 180 km/h (113 mph). At these speeds, polystyrene domains are well above their glass transition temperatures; the thermoplastic elastomers would flow and such a tire would disintegrate. Thermoplastic elastomers can be used for many things such as soles of sneakers, tires of toy cars, adhesives, etc., however.

The search continues. The important feature is the formation of crosslinks in liquid polymers without scrap and such a structure can be achieved in many ways. An Austrian company seems to have succeeded in manufacturing useful polyurethane pneumatic tires by injection molding, first giant tires for slow construction equipment, then tires for fast passenger cars. Even colored tires with comparable properties to conventional black ones can be prepared. How about light green cars with dark green tires? Or something in splendid white or glittering gold? Where is the imagination of designers and customers? Where is the snob appeal?

13 In and Out

*... And the LORD showed him a tree; and
when he cast it into the waters, the
waters were made sweet.*

Exodus 15, 25

13.1 From Moses to Marmelade

The waters were not made "sweet", of course, but their bitterness was
removed. The waters found by the Children of Israel were bitter since they
contained magnesium sulfate, in crystallized form, $MgSO_4 \cdot 7\ H_2O$, known
as Epsom salts or bitter salt. The tree used by Moses acted as ion ex-
changer; it removed magnesium sulfate and other dissolved salts. Scientifically untrained people view this as a miracle, but how else can such trees
get "sweet" water in desolate parts of the world if the trees themselves
cannot act as ion exchangers? Too much magnesium is harmful to plants.

Ion exchangers are nowadays synthesized from polymers; they are used
in great amounts to desalt sea water and brackish water. Many ion
exchangers are based on crosslinked copolymers of styrene with small
amounts of divinylbenzene, $CH_2 = CH - C_6H_4 - CH = CH_2$, which are ret-
roactively reacted to give acidic or basic groups. Polymers with acid
groups act as cation exchangers, those with base groups as anion exchangers. Besides styrene/divinylbenzene copolymers, many other polymers
can be used, for example, phenolic resins, cellulose, and lignin. The last
two polymers are the main components of wood (Table 22). The tree used
by Moses contains acid and base groups bound to cellulose and lignin.
The acid groups remove magnesium ions (Eq. 13-1), the base groups sulfate ions (Eq. 13-2):

$$
\begin{array}{lll}
2\ (poly)^{\ominus}H^{\oplus} & + Mg^{2\oplus} & \rightleftarrows ((poly)^{\ominus})_2 Mg^{2\oplus} + 2\ H^{\oplus} \qquad (13\text{-}1)\\
2\ (poly)^{\oplus}OH^{\ominus} & + SO_4^{2\ominus} & \rightleftarrows ((poly)^{\oplus})_2 SO_4^{2\ominus} + 2\ OH^{\ominus} \qquad (13\text{-}2)
\end{array}
$$

$$
\left.\begin{array}{l}
2\ (poly)^{\ominus}H^{\oplus}\\
2\ (poly)^{\oplus}OH^{\ominus}
\end{array}\right\} + MgSO_4 \rightleftarrows \begin{array}{l}((poly)^{\ominus})_2 Mg^{2\oplus}\\ ((poly)^{\oplus})_2 SO_4^{2\ominus}\end{array} + 2\ H_2O \qquad (13\text{-}3)
$$

Magnesium sulfate is thus exchanged for water. A miracle?

Modern ion exchangers are regenerated by a reversal of processes 13-1 to 13-3 (addition of water; reactions from right to left).

Moses certainly did not regenerate the tree which was quite dead anyway. Nobody cared much about the environment in biblical times, cut trees were not replaced ("... although it is wooded, you shall cut it down ..."; Joshua 17, 18).

Why are polymers used as ion exchangers and not low molar mass anions and cations? Ion exchangers should not be soluble; if they were, water could only be separated by freezing out or distilling off, both very energy-consuming processes. Effective ion exchangers must also possess many acid and base groups. Low and high molar mass chemical compounds with many such groups are however easily soluble in water. The trick is to prevent the dissolution by crosslinking the polymer. Such polymers swell in water but do not dissolve.

Water molecules separate the otherwise tightly bound anions and protons, $(poly)^{\ominus}H^{\oplus}$; the groups dissociate into free polyanions, $(poly)^{\ominus}$ and free protons $H^{\oplus}$ (exact: hydrated protons, $H_3O^{\oplus}$):

$$(poly)^{\ominus}H^{\oplus} \rightleftarrows (poly)^{\ominus} + H^{\oplus} \qquad (13\text{-}4)$$

not dis-   dis-
sociated   sociated

After dissociation, negative groups are lined up along the polymer chain, for example, sulfonic groups in cation exchangers. Anion exchangers dissociate correspondingly into free polycations and hydroxyl anions; their chains are studded with positive groups, for example, ammonium groups.

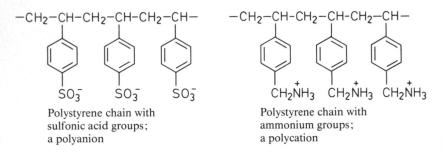

Polystyrene chain with sulfonic acid groups; a polyanion

Polystyrene chain with ammonium groups; a polycation

The many charged groups repel each other, which causes the chains to stretch, the coils to expand, and the specimen to swell. The swelling stops when the retracting forces of the elastic network just balance the tendency of chains to stretch further. The strong swelling exposes additional acid or

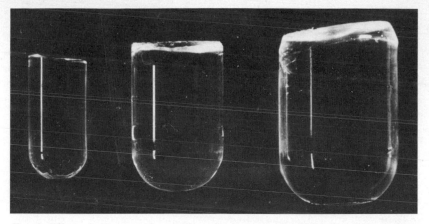

Fig. 49. Effect of solvent on the swelling of polystyrene crosslinked by divinylstyrene. From left to right: unswollen specimen, swelling in cyclohexane, swelling in benzene

base groups in the interior of the ion exchange resin; it is the swelling which creates effective ion exchangers.

The swelling of crosslinked polymers in solvents is not restricted to ion exchangers in water. Even non-charged polymers swell in certain non-aqueous solvents. Swelling is a general phenomenon of chemically or physically crosslinked materials if strong polymer/solvent interactions exist, but which do not destroy the crosslinking sites. Figure 49 shows the swelling of a crosslinked polystyrene by cyclohexane and benzene. Many strong attractions exist between the "aromatic" benzene molecules C_6H_6 and the aromatic phenyl groups $-C_6H_5$ of the polystyrene $\sim\sim(CH_2-CH(C_6H_5))_n\sim\sim$. Cyclohexane C_6H_{12} has no aromatic character and far weaker interactions with polystyrene chains; the swelling in cyclo-hexane is much less than in benzene.

Solvent-swollen polymers often look "solid" although they sometimes consist of more than 99% liquid. Jelly is a good example; it is traditionally made from gelatin, and scientists thus call these "jellies" "gels". Gelatin is a protein obtained from bones or hides.

A typical recipe for a jelly prescribes 1 package gelatin (15 g), 4 table-spoons sugar (50 g), and 1 pint water (450 g). Sugar itself dissolves in water to give a fluid of very low viscosity. It is the 15 grams of gelatin which binds the thirtyfold amount of water and causes the solution to gel.

The amazing binding of water by gelatin is fairly low compared to that of other polymers, though. Acrylonitrile can be grafted to starch and then partially hydrolyzed to give amide and carboxylate groups

$$\underset{\substack{\text{Polyacrylonitrile}\\\text{chains grafted}\\\text{unto starch } (P)}}{(P)-(CH_2-CH)_n-R} \xrightarrow{\;+\,H_2O\;} \underset{\substack{\text{Hydrolysis of}\\\text{nitrile groups}\\\text{to amide groups}}}{(P)-(CH_2-CH)_n-R} \xrightarrow[\substack{-NH_3}]{\;+\,H_2O\;} \underset{\substack{\text{Further hydrolysis}\\\text{of amide groups to}\\\text{carboxylate groups}}}{(P)-(CH_2-CH)_n-R}$$

with the groups CN, $CONH_2$, and COOH respectively below each chain.

Such polymers can bind water to more than 2000 times their weight. Since they are so eager to gobble up water, they got the nice name "super-slurper". Future babies will be happier: diapers will be thinner and the water take-up still greater. But the water uptake has to be very fast since otherwise baby would not be drip-resistant. There is one difference between ion exchangers based on crosslinked polystyrene on one hand and gelatin or grafted starch on the other. Ion exchangers are chemically crosslinked polymers with covalent bonds between the chains (see Chap. 11.1). Gelatin and grafted starch are however physically crosslinked polymers (see Fig. 27). Chemical bonds are much stronger than physical bonds, otherwise they would not be defined as chemical bonds (see Chap. 4.3). It would however be fallacious to conclude that physically crosslinked polymers are always weaker than chemically crosslinked ones.

The strength of the binding between two chains depends not only on the strength of single bonds or contact points but also on the number of such points per chain. Assume a chain with 1000 repeating units. Chemical bond strengths are typically around 300 kJ/mol, physical bond strengths around 4 kJ/mol. If two chains are crosslinked by one chemical bond, then the strength of the link is 300 kJ/(mol chain). If, on the other hand, only 10% of the repeating units exert physical bonds to the other chain, then the combined strength of these physical bonds would be a hundred times 4 kJ/mol or 400 kJ/(mol chain). The physical bond strength between chains would thus be much higher than the chemical bond strength. Unity creates strength, even on molecular levels.

Let us return to gelatin. It is obtained by partial hydrolysis of the α-amino acid chains of collagen, that is, by partial chain scission. Collagen is a triple helix consisting of two equal protein chains and one slightly different one; all with more than 1000 α-amino acid units per chain, for example, 1052 in the so-called α_1 helix of calf skins. Gelatin is prepared from pig hides in the United States and from fresh bones in Europe. Slightly different processes are used for the preparation of gelatin for food and for photographic purposes. The Soviet Union now even produces synthetic caviar from gelatin, since the real stuff became rare thanks to the pollution of rivers.

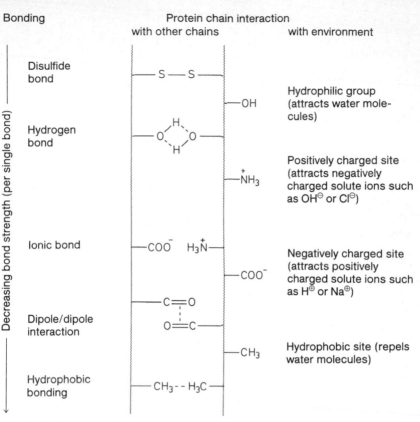

Bonding	Protein chain interaction with other chains	with environment

Disulfide bond — S — S —

OH — Hydrophilic group (attracts water molecules)

Hydrogen bond — O⋯H⋯O —

$\overset{+}{N}H_3$ — Positively charged site (attracts negatively charged solute ions such as $OH^{\ominus}$ or $Cl^{\ominus}$)

Ionic bond — COO⁻   H₃N⁺ —

COO⁻ — Negatively charged site (attracts positively charged solute ions such as $H^{\oplus}$ or $Na^{\oplus}$)

Dipole/dipole interaction — C=O   O=C —

CH₃ — Hydrophobic site (repels water molecules)

Hydrophobic bonding — CH₃ -- H₃C —

Decreasing bond strength (per single bond)

Fig. 50. Binding groups in proteins

Several types of interactions exist between the protein chains of gelatin: chemical bonds such as disulfide bridges, physical bonds like hydrogen bonds, ionic bonds (salt bonds) between groups of opposite charge, interactions between electrical dipoles, and so-called dispersion forces (Fig. 50). All these bonds are present in different amounts; all of them will be influenced differently by various additives. Water H_2O competes with hydroxyl groups — OH for the formation of hydrogen bonds and some — OH/HO — hydrogen bonds between chains are severed by the presence of water. On the other hand, water is hostile to methyl groups and pushes such groups together; it promotes "hydrophobic bonds" (which are entropy driven). The type and extent of such bonds can also be directed by the preparation of gelatin itself. Food gelatin should have properties different from photographic gelatin and carpenters' glue (a strongly

Table 28. Industrial world consumption of water-soluble polymers in 1981 (without Eastern block countries; without starch; without foodstuffs)

Application	Consumption in tons per year		
	Based on natural products	Fully synthetic	Total
Papermaking	4 300 000	600 000	4 900 000
Textiles	400 000	100 000	500 000
Oil production	250 000	30 000	280 000
Flocculation processes	50 000	150 000	200 000
Laundry agents	60 000	20 000	80 000
Paints, lacquers	0	70 000	70 000
Cosmetics	18 000	10 000	28 000
Pharmaceuticals	7 000	5 000	12 000
Total	5 085 000	985 000	6 070 000

degraded gelatin) and different from gelatin used for the encapsulation of pharmaceuticals.

Gelatin belongs to the protein group of polymers, but proteins are not the only group of natural, water-soluble or water-swollen polymers. Another large group is the polysaccharides (polysugars) and their derivatives. Examples of polysaccharides are the amyloses and amylopectins, and also various plant gums such as alginates, gum Arabic, carob, pectin, and agar-agar. Other water-soluble polymers can be produced synthetically from natural polymers by chemical transformations; examples are certain cellulose ethers and carboxymethyl cellulose. Some other water-soluble polymers are fully synthetic; examples are

polyacrylamide $\sim\sim(CH_2-CH(CONH_2))_n\sim\sim$
polyethylene oxide $\sim\sim(CH_2-CH_2-O)_n\sim\sim$.

Water-soluble polymers are used for many different purposes (Table 28): additives in paper-making, coatings for pills, flocculating agents for waste water treatment, bases of creams and lotions in cosmetics. thickeners for latex paints, agents for tertiary oil recovery, blood plasma expanders, and many more.

The application of water-soluble polymers, whether natural or synthetic, follows of course the same chemical and physical principles as the utilization of other polymers. Examples are the pectins which occur in many fruits and which are used as gelling agents for marmelades and other foodstuffs.

Pectins are derivatives of polymerized galactoses (milk sugars). In these poly(α-(1→4)-D-galacturonic acids), all methylol groups $-CH_2OH$ of galactose units are replaced by carboxyl groups $-COOH$; some of the carboxyl groups are furthermore esterified by methanol to methyl ester groups $-COOCH_3$. In certain plants, a small amount of hydroxyl groups $-OH$ is also esterified by acetic acid to acetic acid methyl esters $-OOC-CH_3$. The main chain of pectins may also contain other sugar units besides galacturonic acid units, for example, L-arabinose, fucose, L-rhamnose, and D-galactose units. Pectins are by no means "pure" compounds since they do not consist of only one type of monomeric unit; rather they are copolymers with compositions which change not only from plant to plant but also with the age of the fruit.

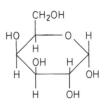

D-Galactose

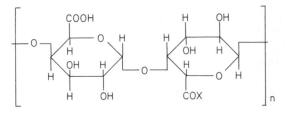

Poly(α-(1→4)-D-galacturonic acid)
(ideal structure; X may be OH or OCH_3)

The differences in the chemical make-up of the various pectins clarify why pectins from citrus plants behave on gelation so differently from those of pine apples. Especially important is the dissociation of carboxyl groups into carboxylate groups and (hydrated) protons

$$\text{P}-COOH \rightleftarrows \text{P}-COO^{\ominus} + H^{\oplus} \tag{13-6}$$

The negatively charged carboxylate groups bind many water molecules. Fewer water molecules are bound by hydrogen bonds to hydroxyl groups (see Fig. 50).

The dissociation equilibrium is shifted to the left if an acid is added to the pectins, for example, lemon juice which contains citric acid. The negatively charged carboxylate groups are transformed into uncharged carboxyl groups. These groups bind less water than carboxylate groups. More bound water is removed from the pectin molecules by the addition of sugar. Household sugar is saccharose (Chap. 6.2) which contains 4 hydroxyl groups per chain unit whereas pectin units possess only two each. Sugar thus has a twice as great chance to bind water than pectin if it is present in equal amounts. There is however always more sugar than pectin. Typical recipes for strawberry jams prescribe at least the same amount of sugar per amount of strawberries. Strawberries contain however less

161

than 10% pectin, so that we always have at least ten times more sugar than pectin and twenty times more hydroxyl groups in sugar than in pectin. No wonder that water molecules desert the pectin and surrender to the sugar. God is always with the stronger batallions.

The pectin macromolecules are thus robbed of their water coats. They feel lonely and associate themselves similar to the protein molecules of foaming spaghetti (Fig. 27). Pectin molecules associate to form a network and the marmelade is ready. How much sugar, pectin, and acid one has to add to fruits to get good marmelades depends very much on the pectin and acid content of fruits. Pine apples, apricots, strawberries, cherries, and rhubarb contain little pectin but are rich in acids. Lemons, oranges, quince, currants and plums, on the other hand, contain both much acid and pectin, especially in the skins.

One word to English purists who insist that only orange jam is true marmelade. The word "marmelade" comes from the Portugese "marmelada" where it denotes a quince jam, not an orange jam. The Portugese word itself can be traced back to the Greek "melimelon" which refers to the fruits of an apple tree grafted on a quince. Thus, what is a true "marmelade"?

13.2 Dreaming of Better Foams

No, we are not talking of airy soap bubbles. Nor do we refer to mountains of foam on dirty lakes and rivers. These foams consist of low molar mass compounds and air. What we like to discuss are rather the plastic foams, otherwise known as foamed plastics or cellular plastics, which are a subgroup of foamed polymers.

A foamed polymer is a polymer which contains dispersed gases in voids throughout the material. The polymers may be thermoplastics, thermosets, or elastomers, and the gases are mainly air, carbon dioxide, or nitrogen. Some foamed polymers have the gases sealed into the voids; they are called closed-cell foams. In other foamed polymers, all voids are interconnected; these foamed polymers are said to be open-cell foams.

Foamed polymers are used for many purposes. We find urea/formaldehyde foams for heat insulation, drinking cups from foamed polystyrene, polyurethane foams for car seats and mattresses, and also cork and wood, as well as bread, cakes, and meringues as food.

Wood? Sure, only the air trapped in voids lets wood float on water (Chap. 11.3); cellulose itself is more dense than water and sinks to the bottom of the lake or river. Bread is a foamed polymer from air and starch, the latter plasticized by water. Bread is a prime example of a (bio)chemically generated foam. By the way, it truly is a synthetic product (though made from natural ingredients) since it does not occur as such in nature (it does not grow on the breadfruit tree).

Another well-known synthetic foodstuff is the meringue, named after the home town of the baker who first beat egg white with air and baked the resulting foam in an oven. The town is Meiringen in the canton of Berne in Switzerland. The baker made his fortune in France, hence Meiringen became "Meringue".

Egg white is an aqueous solution of various more-or-less spherical protein molecules, mainly ovalbumin, globulin, and ovomucin. All these protein molecules exhibit strong intramolecular interactions between the α-amino acid groups within a protein chain (Fig. 50). Some of these physical bonds are broken if the egg white is beaten and the spherical protein molecules denature (Fig. 26) and transform into coil molecules. Formerly intramolecular bound groups are suddenly free and seek new bondage through intermolecular associations (see the foaming spaghettis in Chap. 8.5). The random intermolecular associations create a network which is swollen by water. The resulting "gel" traps the air which the beating incorporated into the egg white. The subsequent heating of the beaten mass causes the entrapped air bubbles to expand and the meringue puffs up. Protein molecules denature even more and new intermolecular associations are formed, making the foam stronger. Water evaporates and the denatured protein "hardens" since it is now below its glass transition temperature.

Many foamed plastics are prepared in exactly the same way, except for the denaturation of molecules of course. Phenol/formaldehyde or melamin/formaldehyde prepolymers are mechanically beaten or vigorously stirred. The resulting foam is hardened by intermolecular chemical crosslinking of the chains. Since the glass transition temperature of the polymers is far above the use temperature, hard foams result. If, on the other hand, natural rubber latex is used instead, a light crosslinking results in soft foams which may be used for cushions or sponges. The true sponge *(Euspongia officinalis),* by the way, has a silica/horn skeleton.

Bread and cakes are puffed up by chemical foaming, not by mechanical beating. Flour is basically starch (Table 18), that is, amylose and amylopectin. Blending the flour with milk or water/milk mixtures creates a more-or-less liquid dough. The flour contains an enzyme (amylase = diastase) which depolymerizes some amylose and amylopectin to glucose at elevated temperatures. Yeasts ferment glucose to alcohol and carbon dioxide which cause the dough to raise. The heat evaporates the water, and since the amylose and amylopectin are no longer plasticized by water the glass transition temperature rises. The starch hardens and the resulting pore structure of breads and cakes is fixed. The alcohol, by the way, evaporates almost completely. But not totally, orthodox anti-alcoholics should thus not eat any bakery goods prepared with yeasts since these goods still contain small amounts of alcohol.

Baking is an art. If the created gas volume is not high enough, doughs will not raise. If the temperature is initially too high, starch will harden before the dough is maximally foamed. Too low a temperature will not create enough gas since the yeast cells do not multiply fast enough to produce sufficient carbon dioxide. The gas production is much more controllable and reproducible if baking powder is used instead of yeasts; the taste of such bakery goods is quite different though (thanks to impurities). Baking powder is a mixture of sodium hydrogen carbonate $NaHCO_3$ (baking soda) and acids such as citric acid, tartaric acid, etc. On heating, the baking soda reacts with the acids

$$NaHCO_3 + R-COOH \rightarrow R-COONa + H_2O + CO_2 \qquad (13\text{-}7)$$

and the carbon dioxide and water then puff up the dough. The same principle is used for the manufacture of foamed plastics.

Hartshorn salt is a mixture of ammonium hydrogen carbonate, NH_4HCO_3, and ammonium carbonate, $(NH_4)_2CO_3$, which decomposes on heating into ammonia, carbon dioxide, and water

$$NH_4HCO_3 \rightarrow NH_3 + H_2O + CO_2 \qquad (13\text{-}8)$$

$$(NH_4)_2CO_3 \rightarrow 2\,NH_3 + H_2O + CO_2 \qquad (13\text{-}9)$$

Besides ammonium hydrogen carbonate, a number of other foaming agents is used, especially those giving nitrogen on thermal decomposition. The reaction of the prepolymers themselves may also produce the foaming gases.

Foamed polymers are especially important in packaging and transportation. They can possess very low densities ("specific gravities") of 0.05 to 0.4 g/mL, much lower than those of glass (ca. 2.55 g/mL) or aluminum (ca. 2.7 g/mL). The low density reduces the dead weight of planes and cars and saves energy. Some foamed polymers are elastic and absorb energy on impact; they are used to package eggs and electronic components.

13.3 All the Things One Packages

Foamed polymers possess large inner surfaces and can thus bind great amounts of effectors. Such an effector ist, for example, hemoglobin which binds and transports oxygen. Hemoglobin immobilized in a polyurethane

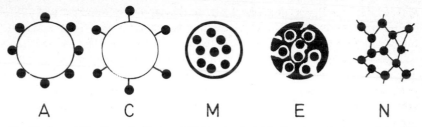

Fig. 51. Immobilization of effectors (●) by physical binding to polymer particles (A), chemical bonding (C), micro-encapsulation (M), encapsulation in a polymer matrix (E), and incorporation into a polymer network (N)

foam takes up oxygen similar to the gills of a fish. Divers can stay under water very long times with this hemo sponge, much longer than with the presently used heavy oxygen cylinders.

The immobilization of hemoglobin in a polyurethane foam is only one of the many possibilities to bind, transport and deliver effectors with the help of polymers. Effectors can be bound physically or chemically to the surface of a polymer, trapped in a polymer particle, encapsulated by polymers, or chemically crosslinked with themselves (Fig. 51). The type of immobilization determines the amounts and rates with which effectors can be set free.

Microencapsulation is probably the most important process. It was discovered in 1953 by scientists of the National Cash Register Company who succeeded in coating small oil droplets with a thin layer of aqueous gelatin, which was then gelled and hardened. The resulting gelatin microcapsules of 1–5000 µm diameter encapsulated the liquid oil.

Gelatin can be replaced by many other polymers. Not only oil droplets but many other liquid or solid materials may be encapsulated, for example, carbon black, dyestuffs, vitamins, bleaching agents, aromas, rat poisons, herbicides, insecticides, pharmaceuticals, fertilizers, and adhesives.

Most curious is probably gasoline in cubes. Gasoline was encapsulated and the microcapsules glued together in cubes. Who needs cubed gasoline? Commandos, according to farsighted military planning! Jeeps of commando troops need gasoline and that is neither easy to get behind enemy lines nor easy to drop from planes. Gasoline engines do not like encapsulated gasoline, however, only liquid fuel without polymer particles. And how does the bold warrior separate the gas from the chaff? By putting the cubes through the wringer ...

Microcapsules are mostly used for carbonless copy papers. These papers do indeed no longer need carbon paper. The microcapsules are glued to the backside of the original form. Writing on the front side breaks the polymer shell and the liberated ink is deposited on the copy. The 600000 tons of such papers which are annually used in the Western World including Japan, contain over 38000 t of dry capsules. The commercial success of its microencapsulation process is said to have prompted the National Cash Register Company to change its name to NCR Corporation and it is rumored that NCR now stands for "No Carbon Required". It may, however, have been the modern trend to short, unintelligible abbreviations; the name of the computer company NBI is supposed to stand for "Nothing But Initials".

Encapsulated material can also be set free by other means than pressure. The shells may be redissolved either physically or chemically. They need not to be destroyed, though, since porous shells allow the encapsulated material to permeate slowly out of the capsule.

In the past, easy soluble fertilizers had to be applied in regular intervals. These fertilizers are however also easily washed out by rain and deposited in rivers and lakes where they may cause eutrophication. A lot of labor could be saved by fertilizing just once for the whole growth period. Microencapsulation of fertilizers may thus be helpful. If the polymer shells are slightly permeable to water, the fertilizer is slowly released over 6–9 months. The same process takes place in the so-called bioorganic fertilization. The inorganic effectors are here encapsulated by the organic materials which slowly decompose and release the fertilizing inorganic compounds.

Encapsulation is especially important for pharmaceuticals. Drugs often taste bitter and are therefore coated with neutral or sweet tasting compounds. More important are three other actions of the encapsulating material: pharmaceuticals should be protected from oxidation by air, the release should be slow, and overdoses should be prevented.

The simple encapsulation of drugs is not an optimal procedure. Neither is the effector applied at its ultimate locus of action nor is the dose delivered regularly. A swallowed pill releases the drug in the stomach; the drug then travels through the blood to all parts of the body. Only a small part arrives at the intended place, say, the liver. Since the amount of drug has to be optimal in the liver, a great overdose has to be applied to the body where it may cause undesired side-effects. Too much medicine is toxic, too little does not help. Since the drug concentration in the liver decreases after a while through the metabolism of the body, new pills have to be taken. This kind of application thus creates maxima and minima in the effector concentration at the organ. The minima are often below, the maxima often above the optimal therapeutic concentrations (Fig. 52). This roller coaster effect can be eliminated if the drug is directly released in constant concentration at the desired location.

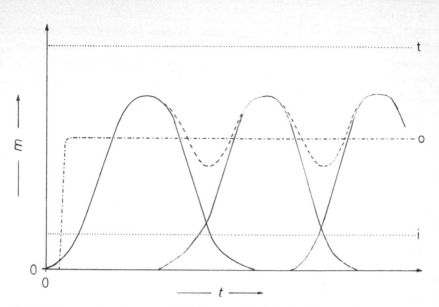

Fig. 52. Mass *m* of an pharmaceutical effector at the desired location as function of time *t* for periodical oral delivery (——). The desirable constant delivery (- · - · - ·) of the optimal amount (O) may be achieved by immobilized systems. t = Toxic amount, i = non-effective amount

New therapeutic systems achieve this goal through effector reservoirs which are surrounded by polymer membranes. So-called transdermal therapeutic systems (TTS) are, for example, placed on the skin as near as possible to the sick organ. Nitroglycerol-TTS, for example, is a coronar therapeutic. Its nitroglycerol reservoir is covered on one side by an impermeable film and on the other side by a membrane with very small pores. The edges of the membrane are coated with a glue which adheres to the skin. The impermeable film, the porous membrane and the glue are all polymers.

The nitroglycerol-TTS is placed on the left side of the chest near the heart. The nitroglycerol which is dissolved or dispersed in a solvent is slowly released through the porous membrane to the skin and finally to the heart itself. Such a transdermal delivery is much more reliable and controllable than oral delivery. Scientists have even developed systems where a constant delivery of an effector is periodically interrupted. It seems also possible to direct the delivery of effectors by small electric currents to electrically conducting polymers which act as valves.

Other research tries to eliminate the undesired side effects of low molar mass pharmaceuticals through their binding to macromolecular carriers.

Such pharmaceutically active polymers should transport the effector through the blood to the place of application where it will be released or become active while still being attached to the polymer. The molar mass of the carrier polymers is however limited by the biological structure of the targeted place of action. Coil-like macromolecules with molar masses in excess of ca. 1000 g/mol cannot penetrate the blood/brain barrier and thus cannot enter the brain and the spinal marrow. Polymers with molar masses over ca. 80000 g/mol cannot enter the tubular epithelium of the kidneys but are slowly excreted through the liver and the gall bladder into the intestines.

Polymer membranes used in pharmaceutical systems are generally membranes with pore diameters of at least 4 nm. Small molecules can, however, permeate through non-porous polymer films, which is an important consideration for packaging films. Polymer chains are not tightly packed in amorphous or partially crystalline polymers; the jumble of chain segments in coil molecules creates certain free spaces (see Fig. 14). These free spaces are regions of atomic dimensions; the overall "free volume" is responsible for the lower density of amorphous polymers as compared to crystalline polymers.

The presence of free volume is probably the most important reason for the ability of gases and vapors to permeate through polymer films. Firstly, the permeating molecules must be soluble in the polymer film; the solubility is the higher the greater the free volume. Little will permeate to the other side of the polymer film before all free space is filled with dissolved gases. The solubilization of permeating gases takes some time and a time lag is therefore observed before the permeating gases exit on the other side of the film. If camphor is sealed into films of 20 μm thickness, it will take 1 day in polyvinylidene chloride before it can be smelled but 92 days in cellulose coated with polyvinylidene chloride and 165 days in films from polyvinyl alcohol.

Secondly, free volume also promotes the diffusion through polymer films. A diffusing gas molecule is almost always bigger than a free space. In order to squeeze through the jungle of chain segments, adjacent segments have to be pushed aside. The required energy is the smaller, the greater the distance between the chain segments, that is, the bigger the free volume.

The permeation of gases and vapors through polymer films is thus controlled by two factors, solubility and diffusion, which together give the permeability coefficient of polymers. Permeability coefficients vary widely: the permeability coefficient of oxygen in silicone rubber is 2 million times higher than that of oxygen in polyacrylonitrile. Typically the higher the oxygen permeability of a polymer, the higher is the permeability for carbon dioxide (Table 29).

Table 29. Permeability coefficient P of oxygen (O_2), carbon dioxide (CO_2), and water vapor (H_2O) in various polymers at $30\,°C$

Polymer	$10^7\ P/(cm^2 \cdot s^{-1})$		
	O_2	CO_2	H_2O
Silicon rubber	605	3 240	40 000
Polystyrene	2.6	10.5	1 200
Polypropylene (isotactic)	2.2	9.2	65
Butyl rubber	1.3	5.2	120
Cellulose 2½ acetate	0.8	2.4	6 800
Polyethylene	0.40	1.8	12
Polyvinyl chloride	0.045	0.16	275
Polyethylene terephthalate	0.035	0.17	175
Polyvinylidene chloride	0.0053	0.029	1.0
Polyacrylonitrile	0.0003	0.0018	300

Plastic bottles for soda pop must have a low permeability for carbon dioxide since otherwise the nice bubbles would disappear after a short time on the shelf. Polyacrylonitrile would be an ideal material, but it was found that the acrylonitrile (monomer!) causes cancer in rats when given in high doses. Proponents of the one-molecule theory speculated that one toxic molecule would be sufficient to cause cancer; ergo, the very small amounts of residual monomer in such bottles would constitute a health hazard. More realistic (and very time-consuming) testing has shown that such adverse health effects do not exist and plastic bottles containing acrylonitrile units are now permitted for the use as foodstuff containers.

The controversy about acrylonitrile-based polymers promoted the use of another polymer, polyethylene terephthalate. This polymer has a relatively high permeation coefficient if it is conventionally processed (Table 29). However, if the polymer is drawn during the bottle-making process, the drawing partially orients the chain segments and they become more tightly packed. The fraction of free volume is reduced and the permeability for carbon dioxide decreases.

Packaging films for fresh fruit, fish, meat and vegetables, on the other hand, require "breathing" polymer films. These films should be highly permeable towards oxygen and water vapor. This can be achieved with plasticized polyvinyl chloride films.

Contact lenses present another problem. Eyes must constantly be supplied with oxygen. The tear liquid on which the contact lens floats is the only provider of oxygen for hard contact lenses from polymethyl methacrylate. Oxygen can however easily permeate soft lenses of silicone rubber or those of poly(hydroxy ethyl methacrylate) = poly(ethylene glycol methacrylate) which is crosslinked with small amounts of ethylene glycol dimethacrylate

$$\begin{array}{ccc}
\text{\tiny{wwww}}CH_2-C(CH_3)\text{\tiny{wwww}} & & \text{\tiny{wwww}}CH_2-C(CH_3)\text{\tiny{wwww}} \\
| & & | \\
COOCH_2CH_2OH & & COOCH_2CH_2OOC \\
& & | \\
& & \text{\tiny{wwww}}C(CH_3)-CH_2\text{\tiny{wwww}}
\end{array}$$

Even paint films for concrete and brick walls have to be permeable for oxygen, carbon dioxide and water. Fresh masonry "breathes" since the mortar has not completely reacted (see Chap. 11.3). Ions, on the other hand, should not be able to permeate through concrete since they may cause corrosion of the reinforcing steel.

13.4 Sticking Together and Holding Apart

In 120 B.C., papyrus was sized with modified wheat starch according to Caius Plinius Secundus (23-74 A.D.) who became known as Pliny the Younger in English speaking countries. Wheat flour was cooked with dilute vinegar (an aqueous solution of acetic acid). The acetic acid hydrolyses the amylose and amylopectin of the wheat starch and short chain polyglucoses with the desired adhesive properties result. Such partially degraded starch can be used to size paper, that is, to make its surface more smooth.

Starch as a glue for paper was known to the Egyptians much more earlier, ca. 3500-4000 B.C. Bone glue, a severely hydrolyzed collagen from bones, hides and hoofs was also known to them about 1500 B.C.

This bone glue or carpenter's glue was *the* glue for thousands of years. Its application was simple: water was added to dry glue and the mixture heated in a water bath. The viscosity of the glue was regulated by adding or evaporating water, that is, by varying the collagen concentration. Bone glue is a solution of short-chain collagen proteins in water; the molar mass of these molecules does not change on dilution.

Bone glue is a solution adhesive just as aqueous solutions of starch or solutions of rubber in organic solvents used for tire repair. Solution adhesives are only effective if they wet the surfaces of the parts (adherents) being glued together. The wettability depends on the critical surface tension; a solution adhesive must possess a lower critical surface tension than the surfaces themselves. Polyvinyl chloride with a critical surface tension of $39 \cdot 10^{-5}$ N/cm and polystyrene with $34 \cdot 10^{-5}$ N/cm can not be wetted

170

by water-based adhesives since water has a surface tension of $72 \cdot 10^{-5}$ N/cm. Not all glues thus work with all surfaces.

The same effect can however be utilized for another purpose: non-wetting also means non-sticking. Animal and vegetable oils and fats exhibit surface tensions of $(20-30) \cdot 10^{-5}$ N/cm and tend to stick to materials with higher surface tensions like metals, glass or enamel. Polytetrafluoroethylene $\sim\sim(CF_2-CF_2)_n\sim\sim$ (Teflon) possesses however a lower surface tension $(18 \cdot 10^{-5}$ N/cm) than oils and fats. Frying pans are therefore coated with Teflon to produce non-sticking surfaces.

Wetting alone does not make a good adhesive since strong bonds have to be formed in order to keep two materials together. Solution adhesives swell the surface of polymers and make the polymer segments flexible. The polymer segments of the adhesive diffuse into this softened layer similar to the self-diffusion of rubber molecules (Chap. 8.4). Both adherent and adhesive must be compatible with each other, of course.

The surfaces of metals and glasses cannot be swollen by solvents, however, and the bonding of their surfaces cannot occur through anchoring polymer molecules of the adhesives in both surfaces. Rather, groups on the surface of the adherent have to react chemically or physically. Glass and metals (with the exception of a few precious metals like gold) are chemically quite different on the surface and in the bulk. Glass surfaces, for example, contain many silanol groups, $-Si-OH$, besides other ones. Such silanol groups may react physically with the adhesive, for example, by forming hydrogen bonds similar to the hydroxyl groups in proteins (Fig. 49) or they may react chemically, for example, by esterification.

The chemical reaction of surface groups is utilized by the so-called polymerization adhesives. Such adhesives are monomers or low molar mass polymers which polymerize, sometimes also with crosslinking. An example is "crazy glue" (Chap. 5.3). The polymerization of the cyanoacrylate monomer of this adhesive not only may be initiated by water vapor $H-O-H$ in air but also by silanol groups $\geqslant Si-OH$ on glass surfaces or by hydroxyl groups $>Fe-OH$ on the surface of iron. The polymerization starts directly on the surface and the polymer chains of the adhesive are firmly anchored to the surface.

Cyanoacrylate adhesives are one-component glues; they do not need a deliberately-added second component for their action. Epoxy glues, on the other hand, are two-component adhesives which need two different components for their chemical reaction. Household epoxy glues normally use the epoxy resin itself as one component and an amine hardener as the second component (which is the one with the fishy smell).

Polymerization adhesives differ also from solution adhesives in one other aspect: the polymerization, and thus the adhesive action, cannot be

171

stopped once it has started. The progressing polymerization generates higher and higher viscosities which cannot be easily lowered by simple dilution with a solvent. The workers of a sand paper factory experienced this phenomenon during the early Fifties. The company was just replacing bone glue by phenol/formaldehyde resins. These resins crosslink during the polymerization which brings about an explosive increase of the viscosity (see Chap. 5.4).

Exactly that happened. The company chemist was absent, and while the workers were on their coffee break, the viscosity got too high and the glue could no longer be coated onto the fabric. Their long experience with bone glue caused the workers to dilute the reaction mixture with water. The startling result let them then call the author who was working in the same company as an elevator operator. Too late: on a distinctly phenol-smelling white broth swam big chunks of gray phenolic resin . . .

What happened? The viscosity increase of the phenolic resin was not only caused by an increased concentration of the resin (the sole reason for viscosity increases in bone glue) but also by an increase of the molar mass through the crosslinking polymerization reaction. Dilution of the reaction mixture with water did not help at all since the phenolic resin is not water-soluble. Only the remaining low molar mass components like phenol are water-soluble. The resulting phase separation created the broth.

A third group of adhesives consists of so-called melt adhesives which are a group of amorphous or partially crystalline polymers applied as a melt above their glass and melt temperatures. Melt adhesives (or hot melts) act through solidification. They are easy to apply but their adhesive strength is not very high. Hot melts are therefore used for temporary adhesion or non-critical applications, for example, for the temporary fastening of inlays in suits before sewing or as labels on containers.

All adhesives, whether melt, solution or polymerization adhesives, are polymers. Modern adhesives are so effective that they are used more and more to join materials, replacing other methods like rivetting or welding. During World War II, the British fighter Mosquito was manufactured from glued veneers. Rotor blades of helicopters and jet fans of planes are now produced from graphite fiber reinforced epoxy resins. Rivets have been replaced by adhesives and such adhesive joints in jumbo jets should survive 90 000 flying hours and 10 million vibration cycles over 30 years. Passenger cars are short-lived consumer goods by comparison. The average car is driven no more than 160 000 km (100 000 miles). The average speed (city and freeway) seldom exceeds 65 km/h (40 miles per hour). A passenger car thus lasts only 2500 actual driving hours. Even a lowly refrigerator does better.

14 Charges and Currents

*"Begin at the beginning," the King said gravely,
"and go on till you come to the end: then stop."*

Lewis Carroll, Alice in Wonderland

14.1 Siemens Inverts Ohm

What wise advice by the king and so fitting for all occasions, especially for this book. We started with the discovery of the strange mechanical properties of rubbers and plastics and we will end with their recently discovered, even stranger properties.

The properties in question are the electrical properties of polymers. The oldest plastics were electrical insulators, the newest ones are either insulators or conductors, even superconductors. The oldest plastics helped found the electrical and electronics industry, the newest ones may revolutionize them.

Celluloid (cellulose nitrate plasticized with camphor) and bakelite (phenolic resin) were revolutionary as substitutes for ivory (Chap. 2) but found much greater applications in other fields. These polymers were especially useful for the young electrical industry which was searching for good electrical insulators during the last years of the nineteenth century. Such electrical insulators were needed for a multitude of applications: as strong solids for cores, cylinders and connectors, as elastic materials for cables and insulating films, and as plastic masses for impregnating items such as magnetic windings.

Materials for these applications were delivered by the young plastics industry. Plastic parts, even complicated ones, were easy to manufacture; fewer parts had to be assembled in a simpler way. An example is the case of a modern telephone. Such a telephone case need not necessarily be made from plastics, one could manufacture it from wood or metal. However, how many steps would be needed for the assembling of wooden parts? And what would be the cost?

Plastics are absolutely necessary for electrical insulations. Classic plastics like phenolic resins, polystyrene, and polyethylene do not conduct electricity. They are such good electrical insulators that the electrical industry used "plastic" and "insulator" as synonyms for many decades.

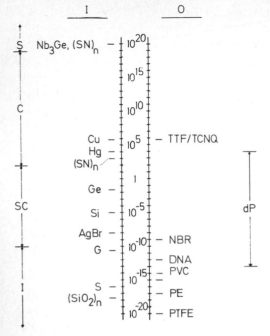

Fig. 53. Electric conductivity of inorganic (I) and organic (O) compounds, measured in S/cm. Triniobium germanide (Nb₃Ge) and poly(thiazyl) (SN)ₙ are superconducting materials at very low temperatures near zero kelvin. The conductivities for conducting (C), semi-conducting (SC) and insulating (I) compounds are given for 20 °C (= 293,16 K = 68° F). Cu = Copper, Hg = mercury, Ge = germanium, Si = silicon, AgBr = silver bromide, G = glass, S = sulfur, (SiO₂)ₙ = quartz, TTF = tetrathiafulvalene, TCNQ = 7,7,8,8-tetracyanoquinodimethane, NBR = nitrile rubber (a copolymer from acrylonitrile and butadiene), DNA = deoxyribonucleic acid, PVC = polyvinyl chloride, PE = polyethylene, PTFE = polytetrafluoroethylene

No longer: plastics can also be electrical conductors and some researchers speculate that only polymers may deliver materials which are superconducting at room temperature.

The new trend may even be observed with the use of physical quantities. The electric resistance, measured in ohms, used to be the prime physical quantity of plastics as far as electric properties were concerned; the electric conductance was given as inverse ohms (mhos). Today, electric conductance seems to be more important than electric resistance; it is now expressed in siemens, the inverse of ohm.

Electrical engineering judges materials according to their conductivity, formerly called "specific electrical conductivity", the ratio of conductance and length. The conductivity is thus a length-related electric conductance. It should not be called "specific" since this term is now reserved for mass-related quantities (example: specific volume = volume/mass).

Conductivities range from 10^{-20} to 10^{20} S/cm, about 40 decades. At one end of the scale are electric insulators with conductivities of 10^{-20} to 10^{-10} S/cm. They are followed by semiconductors (10^{-10} to 10^{2} S/cm) and electric conductors (10^{2}–10^{6} S/cm). The electric conductivity of superconductors is estimated as 10^{20} S/cm (Fig. 53). Superconductivity has sofar only been observed for selected materials at low temperatures; the highest temperature for superconductivity was claimed to be $-148\,°C = -234°$ F for an undisclosed material (as per March 1987).

Such a truly gigantic difference of 40 decades in electric conductivities has never been observed for any other physical property. It is real and not an artefact of our system of physical units. The ratio of 40 decades does not change if, for example, time is measured in days instead of seconds as one can easily check by converting S/cm into basic physical units (1 S/cm $= 100$ $(A^2 \cdot s^3)/kg \cdot m^3)$, see Table A-2).

The difference of 40 decades is truly astronomical. Our universe possesses a diameter of several thousand million light years, that is, about 10^{10} light years. One light year corresponds to $9.46 \cdot 10^{12}$ km $\approx 10^{13}$ km. The diameter of the universe is thus approximately 10^{23} km or 10^{35} nm. The smallest unit of the electric charge is given by the radius of the electron (10^{-12} cm $= 10^{-5}$ nm). The universe thus extends $10^{35}/10^{-5} = 10^{40}$ times the length of the elementary electric charge, which is the same as the ratio of the extremal values of electric conductivity. Coincidence, numerology, or profound meaning?

14.2 Insulators and Xerox

Some years ago it seemed to be obvious why only metals could be electric conductors and why plastics and other organic compounds were insulators instead. Metals are electrically conductive because they possess many electrons which are shared by all the atomic nuclei and can move relatively freely when an electric potential is applied. The conductivities of true metals thus depend relatively little on the chemical nature of the metals: 650000 S/cm for copper, 410000 S/cm for gold, 370000 S/cm for aluminum, and 100000 S/cm for iron.

The situation is quite different for electric insulators where conductivities vary over many decades (see Fig. 53). Such electric insulators may be quartz $(SiO_2)_n$ with a regular network (lattice) of siloxane chains, sulfur with small S_8 rings, long polymeric chains such as in polytetrafluoroethylene $\sim\sim(CF_2 - CF_2)_n\sim\sim$ or polyvinyl chloride $\sim\sim(CH_2 - CHCl)_n\sim\sim$, or double helices as in DNA.

All these compounds bind their atoms through electron pairs (Chapt. 3.3). The electrons of these electron pairs cannot move freely and hence cannot conduct electric currents. Some bonds are however a little bit "polar", that is, the binding electron pairs are not shared to the same amount by both bonded atoms. The electrons are rather a little bit nearer to one atom which becomes a trifle negatively charged. The other atom must therefore become slightly positively charged. The two bonded atoms thus behave as a "dipole" with one negative and one positive end.

An electric field will orient these dipoles in the field direction. In addition, new dipoles are formed, the so-called induced dipoles. The imposed electric field polarizes certain atomic groups since it moves the negatively-charged electrons in one direction and the positively charged atomic nuclei in the other. The electrons and the nuclei do not become completely separated, however. A stronger electric field will cause some electrons to separate from the binding electron pairs. Ions are generated which conduct to some extent the electric current. The small electric conductivity of insulators is thus caused by ions and not by electrons as in the case of metals.

At even higher electric field strengths, more ions are formed. The electric resistance of the material ceases: a breakdown has occured.

The direction of the electron flow changes with frequency in an alternating electric field, that is, at 50 Hz every $(1/50\,s^{-1}) = 0{,}020\,s$. Dipoles try to align themselves in the direction of the electric field. This takes time, and the faster the change in field direction (the higher the frequency), the more difficult it becomes and the alignment of dipoles lags the change in field direction. The greater the time lag, the more energy that is converted into heat and is lost as power.

As often, nature shows a Janus face. Electric insulators should not heat up in an alternating electric field since they would become a bit conducting and would finally show a breakdown. Thus good electric insulators are polymers which neither have permanent nor induced dipoles. Such apolar polymers are polyethylene, polystyrene, and polyisobutylene.

Polar polymers like polyvinyl chloride $\sim\sim(CH_2 - CHCl)_n\sim\sim$ are not useful as electric insulators since their $C - Cl$ dipoles cause the polymer to heat up in an alternating electric field. The heat generated by the field elevates the temperature above the glass transition temperature of PVC. Thus polymers such as PVC become softened and can be easily welded by high frequency fields, which cannot be done with apolar polymers such as polyethylene.

The good electric insulation properties of polymers also cause a not so pleasant nuisance: electrostatic charging. We all have noticed this phenomenon since household goods from plastics seem to attract dust much more easily than those from wood or metals. Worse yet, records collect dust specks, crackle and have to be cleaned before use; all records are made from polymers, mainly today from copolymers of ethylene and vinylacetate.

Either too many or too few electrons are present on insulated or non-grounded surfaces when static electricity is observed. The extra negative or positive charge may be caused by rubbing the surface with other solids, liquids or gases. Gasoline, a mixture of various apolar, low molar mass hydrocarbons (Chap. 7.3), is pumped at gas stations through hoses made from crosslinked nitrile rubber since that rubber is resistant against gasoline. The friction of the gasoline flowing in the hoses generates electrostatic fields of up to several thousand volts per centimeter which may result in a sudden electrical discharge. The discharge can inflame gasoline vapors and cause an explosion. All gasoline hoses are thus surrounded by metal mantles which conduct the surplus negative or positive charges.

Electrostatic charging is always observed if the conductivity is smaller than ca. 10^{-8} S/cm and the relative humidity lower than 70%. The first condition is almost always fulfilled for common plastics, fibers, and elastomers. The second condition is found in hot deserts and during cold northern winters. The relative humidity may drop to 3% on a hot summer day in Arizona's deserts and to 10% indoors in Canadian winters. Touching a metal door knob after walking on a carpet may then cause a mild electric shock.

This electric shock is caused by a sudden discharge of surplus electric charges on the surface of the skin. It got there from the leather or rubber soles of shoes via the clothing. The charge on the soles, in turn, originated from the friction against the carpet. Leather, rubber and synthetic fibers are all insulators and two different insulators always get charged if they are rubbed against each other. The amount and type of charge depends on the rubbing partners and one can arrange all insulators in a so-called triboelectric series:

Positive	Wool
	Polyamides (nylons)
	Cellulose (cotton, rayon, paper)
	Silk
	Polymethyl methacrylate (Plexiglass, Lucite)
	Polyethylene terephthalate (polyester)
	Polyacrylonitrile (acrylic fiber)
	Polyvinyl chloride (PVC)
Negative	Polyethylene

Wool is at the positive end of this series, polyethylene at the negative one. Wool is thus electrostatically charged against acrylic or polyester fibers, for example, if acrylic socks are worn with woolen suits. The pants then stick stubbornly to the socks, especially when the socks are freshly washed. Dirty socks and dusty pants are rarely electrically charged since dirt is polar and surface charges are either not formed or are conducted away. Electric charging may be prevented if pants and socks are treated with a suitable conductive spray.

The unpleasant effects of electrostatic charging may be used, however. Examples are electrostatic paint spraying and the flocking of fabrics to produce velvety surfaces. In both processes, receiving surfaces and sprayed materials are oppositely charged. Unlike things attract each other; all paint particles are thus „magically" attracted by the oppositely charged surfaces. Almost all sprayed paint hits its target, only little is wasted compared to conventional spraying.

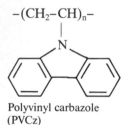

Polyvinyl carbazole
(PVCz)

Electrostatic surface charges also play a role in Xerography. In this copy process, a photo-conducting material is deposited on a metal cylinder. The early photo-conducting material was diarsenic triselenide but polyvinyl carbazole (PVCz) is used nowadays. This polymer absorbs ultraviolet light and, with certain additives, also visible light. PVCz is then charged negatively in the dark by a corona discharge. The image of the object is projected onto the negatively charged PVCz layer. The light coming from the lighter parts of the image discharges the corresponding areas on the PVCz layer whereas the black areas not allowing light transmission cause no discharge. The latent image on the PVCz layer is then sprayed with a powder of a positively-charged, black developer which is coated with a resin. Many developer particles will be attracted to the latent image of the highly negatively-charged dark areas, whereas only few particles will settle on the gray areas and none on the white ones. The new image formed by the positively-charged developer particles is then brought into contact with a negatively charged sheet of paper and the developer is transferred to the sheet of paper. The resulting copy is heated whereupon the resin coating on the developer particles sinters and the picture is fixed.

14.3 How to Convert Insulators into Conductors

If metals conduct electric currents and conventional plastics do not, then the difference, namely freely moving electrons in metals, must be responsible. Conclusion: electrically conducting polymers should result if the chemical structures of polymers would allow to distribute electrons over the whole macromolecule instead of localizing them in discrete atom/atom bonds.

Chemists know in principle how to do that. One has simply to create chemical structures with so-called conjugated double bonds in which double bonds alternate with single bonds. If such structures are also planar, that is, if all bonds of the conjugated system are arranged in a plane, then the double bonds are no longer "localized" according to present chemical doctrine. The alternating double/single bonds are rather smeared over the whole molecule (see also Fig. 19). A simple example (and very special case) for such a chemical structure is the benzene molecule C_6H_6 in which three double bonds formally alternate with three single carbon/carbon bonds. In reality, all electrons of the carbon/carbon bonds are distributed evenly over the whole molecule. The resulting delocalization of double bonds is usually indicated by a circle.

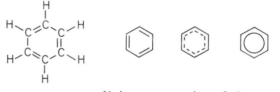

Various representations of a benzene molecule

Chemists also know that graphite possesses a conductivity of ca. 10000 S/cm. Isn't graphite, a polycarbon with platelate structure, a two-dimensional polymer of benzene (Fig. 54)?

Chemists thus synthesized similar polymers which turned out not to be that easy. Poly(p-phenylene), a kind of one-dimensional graphite, was only prepared after many attempts and sophisticated approaches:

(14-1)

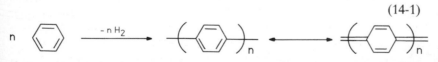

Fig. 54 shows a projection of the poly(p-phenylene) structure onto a section of the graphite structure; the kinship between graphite and poly(p-phenylene) is clearly visible. The two polymers of acetylene

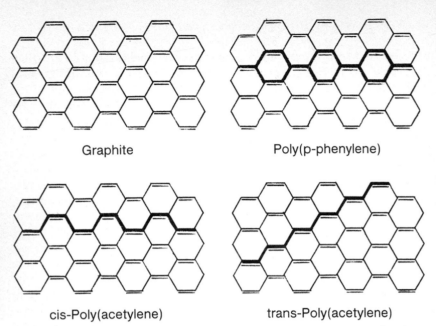

Fig. 54. Sections of a graphite lattice with the corresponding structures of poly(*p*-phenylene), cis-poly(acetylene), and trans-poly(acetylene). Each corner of a hexagon ("ring" in chemical lingo) is occupied by a carbon atom (not shown)

can also be considered kinds of one-dimensional graphite (Fig. 54).

All these gorgeous new chemical structures did not conduct the electric current as expected. They behaved more like electric insulators. Conductivities of cis-polyacetylene and trans-polyacetylene were only 10^{-9} S/cm. But films from these polymers exhibited one very remarkable feature: they glistened like metals.

In hindsight, these disappointingly low conductivities are not so surprising after all. The high electric conductivity of graphite of ca. 10^4 S/cm is observed only in the graphite plane. It is four decades lower perpendicular to the graphite layers. Since poly(*p*-phenylene) and the two polyacetylenes are "one-dimensional graphites", shouldn't the electric conductivity of these materials automatically be much lower than that of graphite?

The unhindered mobility of electrons in a molecule with conjugated double bonds is obviously important; but equally important is the ability of the molecule assembly to transfer charges from one molecule to another. Such charge transfer was observed between certain complex salts of organic compounds, for example, between tetracyanoquinonedimethane (TCNQ) and tetrathiofulvalene (TTF). TTF donates electrons which are accepted by TCNQ. A charge-transfer complex (CT complex) results.

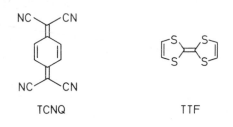

TCNQ   TTF

Charge-transfer complexes do indeed conduct electricity. The conductivity of TCNQ/TTF complexes is 100 S/cm at room temperature and higher than that of copper at 50 K ($= -223\,°C = -369\,°F$).

Organic compounds with such "metallic" conductivities are called "organic metals" or "synthetic metals" ("synmetals"). The name organic metals may be tolerable if one insists that "metal" and "electric conductivity" are synonyms (are they?). But the term *"synthetic* metal"? With a few exceptions, metals are synthesized by man, most are thus synthetic. Back to the future to Chapter 1 . . .

The high electric conductivities of organic metals are only observed for their crystals and only in the directions in which they are stacked. Powdery organic metals show conductivities which are lower by factors of 1000 to 100 000. The crystals of organic metals are also neither mechanically stable nor resistant against oxygen or water. Charge transfer complexes of low molar mass organic compounds can thus not be utilized as electric conductors.

$$-(CH_2-CH)_n-$$

Poly(2-vinylpyridine)

Polymers offer many advantages, however. The complex from poly-(2-vinylpyridine) and iodine exhibits a conductivity of 10^{-3} S/cm. This conductivity is far too low for the application as high-voltage cables in transmission lines. It is however high enough for cathodes of lithium/iodine batteries for implantable pace makers. Such solid state batteries have higher energy densities than the best lead batteries and life times of about ten years. Who would want to carry lead batteries and undergo frequent heart operations?

Much higher electric conductivities are observed with other "doped" polymers. The electric conductivity of polyacetylenes jumps by eleven decades, from 10^{-8} S/cm to 1200 S/cm, when these polymers are doped with arsenic pentafluoride, AsF_5. Most surprisingly, these high conductivities may also be achieved with doped, but non-conjugated polymers. Poly(p-phenylene sulfide), PPS, is a white, non-conducting powder which may be pressed into films having a conductivity of 10^{-16} S/cm. The transparent films become blueish black with the addition of AsF_5 and the conductivity increases to 1 S/cm. These AsF_5-doped poly(p-phenylene sulfides) can be processed as thermoplastics; they are also not attacked by oxygen.

(Poly(p-phenylene
sulfide)

Why these polymers conduct electricity is a big mystery. Some scientists think that the electric conductivity of polyacetylene is caused by "solitons" and that of poly(p-phenylene) by "polarons" (Fig. 55). The soliton is a mathematical concept which received its name in analogy to the solitary waves of water. A solitary wave was first observed by the English engineer John Scott Russell in the nineteenth century. He saw a wave roll along a narrow channel for miles without apparent change of the nature of the wave. This curious stability of the wave is caused by the peculiar movement itself. Solitons are non-linear waves which are described by non-linear mathematical equations.

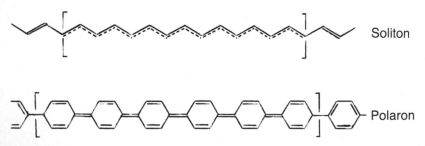

Fig. 55. Schematic representation of a soliton and a polaron. Bond structures change at the square brackets

Electric solitons are described by the same mathematical formalism. The length of an electric soliton is given by the change in chemical bonding (see Fig. 55). A soliton has no magnetic moment but possibly a charge which is only a fraction of the elementary or unit charge of an electron, something which was thought to be impossible some years ago.

Whatever the direction of the physical theory of the electric conductivity of doped polymers, nobody doubts the enormous importance of such materials. Batteries from these polymers would weigh only one tenth of those of lead batteries and occupy only one third of their volumes. Polymer batteries could have ten times the power density of lead batteries and could come in any desired shape, for example, as thin layers under the roof of a car. They will be dry batteries which will not leak or develop toxic fumes. The electrically driven, exhaust-free, non-polluting car may finally come true.

Electrically-conductive polymers should also be of interest to computer manufacturers. Superfast computers, so-called nanocomputers, may be a possibility. Some scientists dream of a "molecular computer" which can store up to 10^{15} bytes of information per cubic centimeter. Utopia? Perhaps not. The realization requires however a lot of research and development ...

We shall not cease from exploration
And at the end of all our exploring
Will be to arrive where we started
And know the place for the first time.

T. S. Eliot, Quartets

Suggested Readings

History

H. Morawetz, "Polymers. The Origins and Growth of a Science," Wiley-Interscience, New York 1985 (comprehensive monograph)

R. B. Seymour and G. A. Stahl, eds., "Genesis of Polymer Science", American Chemical Society, Washington, D. C., 1985 (selected biographies)

R. B. Seymour, ed., "History of Polymer Science and Technology", M. Dekker, New York 1982 (book version of papers in Journal of Macromolecular Science – Chemistry, Volume *A 15* (1981) pp. 1065–1460) (reviews of important polymer applications and concepts, sometimes sketchy)

H. Staudinger, "Arbeitserinnerungen", Dr. A. Hüthig Verlag, Heidelberg 1961 (personal memoirs in German)

C. Priesner, "H. Staudinger, H. Mark and K. H. Meyer – Thesen zur Größe und Struktur der Makromoleküle", Verlag Chemie, Weinheim 1980 (documented history of the emergence of the macromolecular concept; in German)

Y. Furukawa, "Staudinger, Carothers, and the Emergence of Macromolecular Chemistry", Ph. D. Thesis, University of Oklahoma, 1983; University Microfilms, Ann Arbor 1984, 83-24888

R. Olby, "The Macromolecular Concept and the Origin of Molecular Biology", Journal of Chemical Education *47* (1970) 168

J. D. Watson, "The Double Helix. A Personal Account of the Discovery of the Structure of DNA", Atheneum, New York 1968

R. Olby, "The Path to the Double Helix", University of Washington Press, Seattle 1974

F. M. McMillan, "The Chain Straighteners: Fruitful Innovation. The Discovery of Linear and Stereoregular Polymers", The MacMillan Press, London 1981

R. Friedel, "Pioneer Plastic. The Making and Selling of Celluloid". The University of Wisconsin Press, Madison, Wisconsin 1983

E. Rochow, "Silicon and Silicones", Springer, Heidelberg New York 1987

Introductory Textbooks

J. M. G. Cowie, "Polymers: Chemistry and Physics of Modern Materials". Intext Educational Publishers, New York 1974

F. W. Billmeyer, "Textbook of Polymer Science", J. Wiley-Interscience, New York, 3rd ed. 1984

Comprehensive Textbooks

H.-G. Elias, "Macromolecules", Plenum Press, New York, 2nd ed. 1984 (2 vols.)

Appendix

Table A-1. SI Prefixes and common names of numbers

The International System of Units (Systéme International des Unités, SI System) regulates prefixes and symbols for physical units. Such an international agreement is urgently required since words for numbers may have different meanings in different languages or countries. In England, a billion is a number with 12 digits before the comma (the decimal point in America), in the United States, it represents a number with 9 digits. In American financial circles, the symbol M is often used for "thousand" and the symbol MM for "million" whereas the symbol M represents 1 million in the SI system. In international conventions, symbols for units are always written behind numbers. \$7 MM thus becomes 7 M\$ (seven megadollar) in the SI System.

Factor	SI System		Common Name	
	Prefix	Symbol	England and Germany	United States, Soviet Union, France
10^{24}	–	–	Quadrillion	Septillion
10^{21}	–	–	Trilliard	Sextillion
10^{18}	Exa	E	Trillion	Quintillion
10^{15}	Peta	P	Billiard	Quadrillion
10^{12}	Tera	T	Billion	Trillion
10^{9}	Giga	G	Milliard	Billion
10^{6}	Mega	M	Million	Million
10^{3}	Kilo	k	Thousand	Thousand
10^{2}	Hecto	h	Hundred	Hundred
10	Deca	da	Ten	Ten
10^{-1}	Deci	d	One tenth	One tenth
10^{-2}	Centi	c	One hundredth	One hundredth
10^{-3}	Milli	m	One thousandth	One thousandth
10^{-6}	Micro	μ	One millionth	One millionth
10^{-9}	Nano	n	One milliardth	One billionth
10^{-12}	Pico	p	One billionth	One trillionth
10^{-15}	Femto	f	One billiardth	One quadrillionth
10^{-18}	Atto	a	One trillionth	One quintillionth

Table A-2. Physical units in the SI system

Physical quantity	Physical unit	
	Name	Symbol
Basic quantities		
Length	meter	m
Mass	kilogram	kg*
Time	second	s
Electric current	ampere	A
Thermodynamic temperature	kelvin	K
Luminous intensity	candela	cd
Amount of substance	mole	mol
Additional quantities		
Plane angle	radian	rad
Solid angle	steradian	sr
Derived quantities		
Force	newton	$N = kg \cdot m \cdot s^{-2}$
Energy, work, heat	joule	$J = kg \cdot m^2 \cdot s^{-2}$
Power, energy flux	watt	$W = kg \cdot m^2 \cdot s^{-3}$
Pressure, stress	pascal	$Pa = kg \cdot m^{-1} \cdot s^{-2}$
Frequency	hertz	$Hz = s^{-1}$
Electric charge	coulomb	$C = A \cdot s$
Electric potential difference	volt	$V = kg \cdot m^2 \cdot s^{-3} \cdot A^{-1}$
Electric resistance	ohm	$\Omega = kg \cdot m^2 \cdot s^{-3} \cdot A^{-2}$
Electric conductance	siemens	$S = kg^{-1} \cdot m^{-2} \cdot s^3 \cdot A^2$
Electric capacitance	farad	$F = kg^{-1} \cdot m^{-2} \cdot s^4 \cdot A^2$
Relative permittivity	—	1
Magnetic flux	weber	$Wb = kg \cdot m^2 \cdot s^{-2} \cdot A^{-1}$
Inductance	henry	$H = kg \cdot m^2 \cdot s^{-2} \cdot A^{-2}$
Magnetic flux density	tesla	$T = kg \cdot s^{-2} \cdot A^{-1}$
Luminous flux	lumen	$lm = cd \cdot sr$
Illumination	lux	$lx = cd \cdot sr \cdot m^{-2}$
Radioactivity	becquerel	$Bq = s^{-1}$
Absorbed dose of radiation	gray	$Gy = m^2 \cdot s^{-2}$

* Note: The "ton" = 1000 kg is tolerated as a physical unit although it is not an SI unit. It should not be confused with either "long ton" or "short ton".

Table A-3. Conversion of some Anglo-Saxon units (often rounded)

American measures of lengths

1 mile	= 1609.344 m =	1.609344 km
1 yard	=	0.9144 m = 91.44 cm
1 foot	=	0.3048 m = 30.48 cm
1 inch	=	2.54 cm
1 mil	=	0.00254 cm

American measures of areas

1 square mile	=	2.590 km^2
1 acre	= 4047 m^2	
1 square yard	=	0.8361 m^2
1 square foot	= 929 cm^2	
1 square inch	=	6.542 cm^2

American measures for volumes (note: 1 liter = 1 dm^3)

1 US barrel	= 119 L	
1 US barrel petroleum	= 158.97 L = 42 US gallons (liquid)	
1 gallon (US liquid)	=	3.7852 L = 4 quarts
1 quart	=	0.9463 L = 2 pints
1 pint	=	0.4732 L = 2 cups
1 cup	=	0.2366 L = 8 fluid ounces
1 fluid ounce	=	0.02957 L = 2 tablespoons
1 tablespoon	=	0.01479 L = 3 teaspoons
1 teaspoon	=	0.00493 L

English measures for volumes

1 Imperial barrel	= 163.6 L	
1 Imperial gallon	=	4.546 L
1 ounce (British liquid)	=	28.413 mL

Other English and American measures

1 long ton (UK)	= 1016.05 kg	
1 short ton (US)	=	907.185 kg
1 hundred weight (UK)	=	50.8023 kg
1 short hundred weight	=	45.3592 kg
1 slug	=	14.59 kg
1 stone	=	6.35 kg
1 pound (avoirdupois) (lb)	=	453.6 g
1 pound (apothekers)	=	373.2 g

Table A-4. Symbols, names, atomic numbers (in parentheses), and relative atomic masses of chemical elements (Commission on Atomic Weights and Isotopic Abundances, International Union for Pure and Applied Chemistry, 1985)

Most chemical elements are mixtures of various isotopes; their relative atomic masses thus depend on their origin in nature or on their synthesis. An exception are chemical elements composed of a single isotope; these are marked with *

Ac	Actinium (89)	
Ag	Silver (argentum) (47)	107.8682
Al	Aluminum* (13)	26.981539
Am	Americum (95)	
Ar	Argon (18)	39.848
As	Arsenic* (33)	74.92159
At	Astatine (85)	
Au	Gold* (aurum) (79)	196.96654
B	Boron (5)	10.811
Ba	Barium (56)	137.327
Be	Beryllium* (4)	9.012182
Bi	Bismuth* (83)	208.98037
Bk	Berkelium (97)	
Br	Bromine (35)	79.904
C	Carbon (6)	12.011
Ca	Calcium (20)	40.078
Cd	Cadmium (48)	112.411
Ce	Cerium (58)	140.115
Cf	Californium (98)	
Cl	Chlorine (17)	35.4527
Cm	Curium (96)	
Co	Cobalt* (27)	58.93320
Cr	Chrome (24)	51.9961
Cs	Cesium* (55)	132.90543
Cu	Copper (cuprum) (29)	63.546
Dy	Dysprosium (66)	162.50
Er	Erbium (68)	167.26
Es	Einsteinium (99)	
Eu	Europium (63)	151.965
F	Fluorine* (9)	18.9984032
Fe	Iron (ferrum) (26)	55.847
Fm	Fermium (100)	
Fr	Francium (87)	
Ga	Gallium (31)	69.723
Gd	Gadolinium (64)	157.25
Ge	Germanium (32)	72.61

H	Hydrogen (1)	1.00794
He	Helium (2)	4.002602
Hf	Hafnium (72)	178.49
Hg	Mercury (hydrargyrum) (80)	200.59
Ho	Holmium* (67)	164.93032
I	Iodine* (53)	126.90447
In	Indium (49)	114.82
Ir	Iridium (77)	192.22
K	Potassium (kalium) (19)	39.0983
Kr	Krypton (36)	83.80
La	Lanthanum (57)	138.9055
Li	Lithium (3)	6.941
Lr	Lawrencium (103)	
Lu	Lutetium (71)	174.967
Md	Mendelevium (101)	
Mg	Magnesium (12)	24.3050
Mn	Manganese* (25)	54.93805
Mo	Molybdenum (42)	95.94
N	Nitrogen (7)	14.00674
Na	Sodium (natrium)* (11)	22.989768
Nb	Niobium* (41)	92.90638
Nd	Neodymium (60)	144.24
Ne	Neon (10)	20.179
Ni	Nickel (28)	58.69
No	Nobelium (102)	
Np	Neptunium (93)	
O	Oxygen (8)	15.9994
Os	Osmium (76)	190.2
P	Phosphorus* (15)	30.973762
Pa	Protactinium (91)	
Pb	Lead (plumbum) (82)	207.2
Pd	Palladium (46)	106.42
Pm	Promethium (61)	

Po	Polonium (84)	
Pr	Praseodymium (59)	140.90765
Pt	Platinum (78)	195.08
Pu	Plutonium (94)	
Ra	Radium (88)	
Rb	Rubidium (37)	85.4678
Re	Rhenium* (75)	186.207
Rh	Rhodium (45)	102.90550
Rn	Radon (86)	
Ru	Ruthenium (44)	101.07
S	Sulfur (16)	32.066
Sb	Antimony (51) (stibium)	121.75
Sc	Scandium* (21)	44.955910
Se	Selenium (34)	78.96
Si	Silicon (14)	28.0855
Sm	Samarium (62)	150.36
Sn	Tin (stannum) (50)	118.710
Sr	Strontium (38)	87.62
Ta	Tantalum (73)	180.9479
Tb	Terbium* (55)	158.92534
Tc	Technetium (43)	

Te	Tellurium (52)	127.60
Th	Thorium* (90)	232.0381
Ti	Titanium (22)	47.88
Tl	Thallium (81)	204.3833
Tm	Thulium (69)	168.93421
U	Uranium (92)	238.0289
Unh	Unnilhexium (106)	
Unp	Unnilpentium (105)	
Unq	Unnilquadium (104)	
Uns	Unnilseptium (107)	
V	Vanadium* (23)	50.9415
W	Tungsten (74) (wolfram)	183.85
Xe	Xenon (54)	131.29
Y	Yttrium (39)	88.90585
Yb	Ytterbium (70)	173.04
Zn	Zinc (30)	65.39
Zr	Zirconium (40)	91.224

Table A-5. Some trade names and trivial names of polymers

ABS	Generic name for plastics from acrylonitrile, butadiene and styrene units
Acetate	Generic name for fibers from cellulose (2½) acetate
Acrilan	Trade name for fibers from polyacrylonitrile (Chemstrand)
Acryl	Generic name for fibers with at least 85% acrylonitrile units
Alcantara	Synthetic suede from polyester fibers in a polyurethane matrix
Araldit	Epoxy resins (Ciba-Geigy)
Aramide	Generic name for fibers from aromatic polyamides
Bakelite	Thermoset from phenol and formaldehyde (Bakelite Corp.)
Balata	Natural trans-1.4-polyisoprene
Cellophane	Films from regenerated cellulose (Kalle)
Celluloid	Cellulosenitrate plasticized with camphor
Dacron	Fibers from polyethylene terephthalate (DuPont)
Diolen	Fibers from polyethylene terephthalate (Vereinigte Glanzstoff)
Dralon	Fiber from polyacrylonitrile (Bayer)
Grilen	Fiber from polyethylene terephthalate (Ems Chemistry)
Guttapercha	Natural trans-1.4-polyisoprene
Hostalen	Polyethylene thermoplastic (Hoechst)
Hostalen PP	Polypropylene thermoplastic (Hoechst)
Hostalit	Polyvinyl chloride thermoplastic (Hoechst)
Hostaphan	Films from polyethylene terephthalate (Kalle)
Igelit	Polyvinyl chloride thermoplastic (BASF)
Lucite	Polymethyl methacrylate thermoplastic (DuPont)
Lupolen	Polyethylene thermoplastic (BASF)
Luran	Thermoplastic from styrene and acrylonitrile (BASF)
Lustron	Thermoplastic polystyrene (Monsanto)
Marlex	Thermoplastic polyethylene (Phillips Chemical)
Moltopren	Cellular polymer from polyurethanes (Bayer)
Mylar	Films from polyethylene terephthalate (DuPont)
Natural rubber	cis-1.4-Polyisoprene
Neoprene	Elastomeric polymers and copolymers from chloroprene (DuPont)
Nylon	Generic name for polyamides
Orlon	Fiber from polyacrylonitrile (DuPont)
Perlon	Fiber from polycaprolactam (Bayer)
Plexiglas	Thermoplastic polymethyl methacrylate (Rohm and Haas)
Polyester	Generic name for fibers from polyesters with at least 85% terephthalic acid and ethylene glycol units
Polyester, unsaturated	Thermosets from maleic acid/ethylene glycol polymers (or similar compounds) which are crosslinked with e. g. polystyrene

Qiana	Polyamide from trans,trans-diaminodicyclohexylmethane and dodecanedicarboxylic acid or sebacic acid (fiber from DuPont)
Rayon	Generic name for fibers from regenerated cellulose
Saran	Films and fibers from polymers with at least 80% vinylidene chloride units (Dow Chemical)
Silicone	Generic name for polymers with a siloxane chain
Spandex	Generic name for elastic fibers from polymers with at least 85% segmented polyurethane
Styrofoam	Cellular plastic from polystyrene (Dow Chemical)
Styron	Polystyrene thermoplastic (Dow Chemical)
Styropor	Cellular thermoplastic from polystyrene (BASF)
Teflon	Thermoplastic polytetrafluoroethylene (or other fluorinated polymers if letters follow the word Teflon) (DuPont)
Terital	Fibers from polyethylene terephthalate (Montedison)
Terlenka	Fibers from polyethylene terephthalate (AKU)
Terylene	Fibers from polyethylene terephthalate (ICI)
Trevira	Fibers from polyethylene terephthalate (Hoechst)
Triacetate	Generic name for fibers from cellulose triacetate
Tricel	Fibers from cellulose triacetate (British Celanese)
Ultrasuede	Synthetic suede (see Alcantara)
Vestolen	Various polyamides (Huels)
Vestolit	Polyvinyl chloride (Huels)
Viscose	Generic name for fibers from regenerated cellulose
Zytel	Various aliphatic polyamides (DuPont)

Subject Index